Science and Engineering in Ireland in 1798: a time of revolution

Science and Engineering in Ireland in 1798: a time of revolution

Edited by
Patrick N. Wyse Jackson

Proceedings of a symposium organised by the National Committee for the History and Philosophy of Science, 4th November 1998.

Royal Irish Academy
2000

ISBN 1 874045 77 1

First published 2000 by the Royal Irish Academy, 19 Dawson Street, Dublin 2, Ireland.
Printed by Colour Books, Baldoyle, Dublin 13.

Cover: View of Fair Head on the north Antrim coast [see page 36].

Contents

Contributors

Peter J. Bowler, School of Anthropological Studies, The Queen's University of Belfast.

D. Thorburn Burns, Department of Analytical Chemistry, The Queen's University of Belfast.

William J. Davis, Department of Chemistry, Trinity College, Dublin.

James C.I. Dooge, Centre for Water Resources Research, University College Dublin, Dublin.

Patrick N. Wyse Jackson, Department of Geology, Trinity College, Dublin.

Acknowledgements

The National Committee for the History and Philosophy of Science is grateful to the 1798 Commemorative Committee of the Irish Government, which funded the publication of this book, and for the assistance of Ann Kearney of the Taoiseach's Office. It is also grateful to Veronica Barker of the Royal Irish Academy for her diligent assistance in organising the one-day meeting on which it is based.

Introduction

Science in the latter decades of the 18th century was undergoing major theoretical alteration. The New Chemistry of the French and Dalton's Atomic Theory were finding vogue amongst the chemists of the day. Geology was emerging from the limited discipline of German-led mineralogists to become a holistic science of the Earth. The theories of the Edinburgh men of science championed by James Hutton were gaining ground and acceptance. Biology was to undergo its major theoretical advances in the middle of the next century; nevertheless the revolutionary and evolutionary ideas, in both senses, of Charles Darwin were heralded by his grandfather Erasmus. It was the age of the expansion of the canals, which gave the much impetus to the Industrial Revolution in later Victorian Britain. Although the level of industrial expansion was low in Ireland, when compared with mainland Britain, the expansion of the Irish canal waterways was important in opening up Ireland economically.

All of these scientific and engineering advances were taking place against the backdrop of social and political change, particularly in North America and France. The colonialists in the American colonies gained independence from Britain. The French Revolution swept away the landed classes and aristocracy which were replaced by strong-minded idealists who established a democracy highly regarded by many. It was natural that these events should have strongly influenced other nations. In Ireland, reactionaries and those disaffected with the corrupt Dublin Parliament and English domination saw their goals mirrored in the French Republic. The underground United Irishmen turned to France for assistance in their planned rebellion. Foul weather and informants did not help their cause: the bulk of the planned French invasion fleet could not land and the Directorate of the United Irishmen was betrayed to the authorities in Dublin Castle.

The subsequent Act of Union removed from Dublin much power, but in terms of scientific endeavour the work of the Royal Irish Academy, the Royal Dublin Society, Trinity College Dublin, and other local institutions continued to impress well into the nineteenth century. The period 1800–1850 is regarded by many historians of science as being a 'golden age' in terms of scientific expansion and discovery. However, one must go back to the roots of this expansion, to the mid- to late-1700s to understand the cultural and scientific background to this development. The growth of these roots in Ireland coincided with large-scale, country-wide social unrest that culminated in the ill-fated 1798 rebellion. At the same time another, less well-known and appreciated intellectual 'rebellion' was taking place in Ireland and Britain amongst the scientists and engineers.

In this volume Peter Bowler discusses the pre-Darwinian advances in biology; William Davis and Thornburn Burns recall the careers of the exiled chemists W.J. MacNeven, a leading United Irishman, and J.P. Emmet, the son of another. Patrick Wyse Jackson examines the debate pertaining to the origins and nature of some igneous rocks; a debate that centred on northeast County Antrim. Jim Dooge writes on the expansion of civil engineering in Ireland and gives a lively account of the major players in this area.

During the bicentenary of the 1798 rebellion much attention was paid to the military and historical events of the period, and the work of contemporary scientists was largely ignored by commentators. This volume seeks to reddress the balance between history and the history of science and to place on record the achievements of Irish scientists in the late 1700s.

Patrick N. Wyse Jackson
Rathmines, Dublin
26th November 1999

Biology in an Age of Revolution

Peter J. Bowler

Our theme in this volume is the involvement of science and technical innovation in 'revolutionary' activity — specifically in the context of Ireland in 1798, but also, I hope, more generally. It is obvious that scientists — like anyone else — can be caught up in the events of a political revolution. Other papers below show how several scientists were forced to leave Ireland because of their own or their family's involvement in the events of 1798. But I want to stress that the link between science and revolutionary beliefs and actions may be more than coincidental. Historians of science are increasingly conscious of the social context within which science develops, and hence of the cultural, religious and ideological values that can be associated with scientific ideas and with technical innovations. Those who invent new machines or processes usually do so in the hope of having some impact on the economy, and hence on society. They may come from a social background that places a stronger than average emphasis on making use of one's talents in this way. But the same factors apply to those who make theoretical innovations even within what may seem like 'pure' areas of science — indeed the same people were often involved because they made no clear distinction between pure and applied knowledge. One can change the world by changing the way people think about it as well as by manipulating it directly.

The use of the term 'scientific revolution' to denote some of the greatest advances in scientific thinking reminds us that the development of science itself can undergo spurts of sudden activity.[1] It also symbolises the fact that new scientific theories can be perceived as agents capable of transforming culture and society. In my own field, the term 'Darwinian Revolution' highlights the ability of a scientific theory to encapsulate radical ideas that challenge traditional religious beliefs — and in so doing transform both moral and political values. Indeed, I have argued that the implications of Charles Darwin's ideas were so radical that their full impact was delayed into the twentieth century.[2] There can be little doubt that Darwin and his followers were acutely conscious of the revolutionary potential of what they were doing — they were not ivory-tower academics studying nature for its own sake and leaving others to worry about the consequences. When scientists participate in these great events, they do so not as the remote specialists of popular caricature but as human beings who are well aware of the radical implications of the ideas and techniques they are proposing.

My intention in the next few pages is to walk a tightrope of connections between Ireland in 1798 and the broader social and intellectual revolutions with which science has been associated, specifically the Industrial Revolution and the early phases of the Darwinian Revolution. Scientific and technical innovations were at the heart of these transformations, and those who promoted the new ideas were using them quite deliberately to change the world they lived in. They were not, however, necessarily revolutionaries in their local political context, and the case of Ireland shows how local political realities can mask the underlying revolutionary implications of an individual's beliefs. As modern historians are increasingly aware, the local context almost always dictates the way ideas are understood and exploited.

My starting point in 1798 is the events surrounding the French landing in County Mayo and the subsequent uprising. Relatively nearby at Edgeworthstown, County Longford, Richard Lovell Edgeworth had raised a company of yeomen. They marched to Longford, but arrived after the French had been defeated — although Edgeworth then found himself threatened by the Longford Protestants as a suspected spy.[3] As a landlord, Edgeworth was against the revolution, but he was an enlightened thinker who opposed the extremism of many of his fellow landholders. His yeomanry included Catholics as well as Protestants, and the rebels spared his house when they occupied the grounds before their defeat. Here we see the local context at work: Edgeworth wanted social change through enlightened leadership in social and economic affairs, but he could not go along with those who would undermine the whole fabric of existing society.

In the broader context of European history, Edgeworth was an innovator linked to the radicals who were already transforming the economy and undermining the old social hierachy. On his Irish estates he encouraged land improvements through drainage. He was fascinated by mechanical inventions and was encouraged by his contacts in England to publish a book on the subject.[4] Those contacts included the members of the Lunar Society, which brought together the architects of the Industrial Revolution, innovators such as James Watt and Matthew Boulton, whose steam engines were the driving force of technical progress. These were all thinkers with radical ideas about the potential for an innovative elite to transform society through invention but also by challenging the political philosophy that supported the old landed interests of Britain. The middle classes were starting to demand a political role commensurate with their growing economic power. In this sense, Edgeworth's position as an Irish landlord was anomalous — his local interests forced him to adopt a conservative role as an opponent of popular revolution, whatever his desires for social reform in a wider context. In another sense, however, he was typical — the middle-class innovators wanted to restructure the social hierarchy, not abolish it altogether.

Edgeworth corresponded most frequently with Erasmus Darwin, a key figure linking the interest in technical innovation with theoretical speculations designed to undermine the traditional vision of nature.[5] From the letters recorded in his *Memoirs*, we know that Edgeworth read and enjoyed Darwin's poems, including *The Botanic Garden*, and also his *Zoonomia*, published in 1790–94.[6] Here Erasmus Darwin articulated his own theory of organic evolution, anticipating at least in outline the ideas more fully developed by his grandson, Charles. I want to stress the radical, even revolutionary, nature of these early evolutionary speculations. They may not have had the technical sophistication of Charles Darwin's theory of natural selection, but they were a direct challenge to the religious and the social orthodoxy of the time. Traditionalists pictured nature as a static hierarchy designed by God, with the aim of encouraging everyone to accept their position in an unchangeable social order. Radicals such as Darwin now portrayed the universe as a dynamic system, with the efforts and innovations of individual organisms serving as the driving force of progress — just as the technical innovators were transforming the old economy. At the same time, Darwin's position as one of the country's leading physicians allowed him to comment with authority on the latest ideas on the nature of life.[7] For him, living things were active, creative beings with the power to shape their own — and indirectly their descendants' — lives. An important section of William Paley's *Natural Theology* of 1802, that classic exposition of the view that nature is a static, divine artefact, was written to rebut Erasmus Darwin's theories on the ability of living things to take charge of their own destinies.[8]

Paley's static model of nature was central to the conservative orthodoxy which dominated British science and medicine, especially at Oxford, Cambridge and the Royal Colleges of Physicians and of Surgeons. The need for social stability was preached by appealing to a static world created by God in which every species (and hence by implication every creature) had its alloted position. This world view was finally overthrown by Charles Darwin's theory of evolution in the 1860s, and many commentators on the Darwinian Revolution have noted the extent to which evolutionism expressed the ideology of the now dominant middle-class entrepreneurs and innovators. Until recently, it was assumed that the static world view went largely unchallenged in early nineteenth-century science, but thanks to the work of Adrian Desmond we now know that there was an extensive network of support for radical evolutionism in the early decades of the century. Desmond shows how Erasmus Darwin's speculations were replaced by newer evolutionary ideas emanating from revolutionary France.[9] The writings of J.B. Lamarck — whose views on the active powers of living organisms closely resembled Erasmus Darwin's — and Etienne Geoffroy Saint Hilaire were taken up by radical thinkers, especially in the new private medical schools, as a means of undermining the intellectual authority of the establishment. These were middle- and lower-class students and lecturers who demanded a reform of the autocratic governance of the medical profession — and were also often political radicals on a wider scale. That leading organ of the modern medical establishment, *The Lancet*, began its career in 1823 as a radical journal published by one of the most articulate reformers, Thomas Wakley.

In the short term the efforts of the more radical reformers were in vain. The conservatives closed ranks and were able to brand evolutionary theorising as an atheistic and hence subversive activity. Charles Darwin delayed the publication of his own theory, conceived in the late 1830s, because he saw how clearly the general idea of evolution had been labelled as revolutionary. But long before the Origin of Species was published, pressure from the increasingly powerful middle classes had begun to soften the attitude of instinctive hostility towards the idea of progress and hence the theory of evolution. The Edinburgh publisher Robert Chambers, a leading campaigner for middle-class interests, published (anonymously) his *Vestiges of the Natural History of Creation* in 1844. Its vision of nature as designed to progress in accordance with divinely-created laws was howled down by conservatives but was lapped up by ordinary readers, who saw it as a model for social progress.[10] Many historians now believe that *Vestiges* helped to shape the climate of opinion within which Darwin's *Origin of Species* was assimilated fifteen years later. Darwin's theory of natural selection, with its emphasis on the individual who succeeded or failed in the struggle for existence, reflected the ideology of *laissez-faire* individualism, especially when it was interpreted in terms of Chambers' emphasis on progress. In this sense, the Darwinian Revolution was the direct heir to the Industrial Revolution.

The events of the later nineteenth century thus brought to culmination a series of revolutionary scientific and social developments initiated in the last part of the previous century. In the heady days of the 1790s and subsequent decades, radical scientists and innovators had already established a clear link between pressure for social reform, the drive for mechanical invention, and theories of evolutionary biology. In the context of Ireland in 1798, Richard Edgeworth appears as a conservative figure, a landlord who supported the efforts of the government to crush the rebellion. Yet in the broader context he was a radical whose ideas linked in with those of innovators who were transforming the British economy in a way that could only be accommodated if there were major changes in the way society was governed. Their new view of society in turn demanded a new, evolutionary view of nature itself. It may seem a long way from Edgeworth's more enlightened view of how to run his Irish estates to the evolutionary theories of Erasmus and Charles Darwin, but I hope that this brief survey has shown the importance of modern historians' efforts to situate science in its social context. We all know the power of technical innovation to transform a society, but scientific theories themselves can have a similar, if less direct, function. In the decades around 1800, those who were benefitting from the economic developments of industrialisation were often those whose vision of nature encompassed models of change and development anathema to the politically orthodox. Science can be a revolutionary activity — and I hope that the rest of this volume will help to demonstrate this point in more detail.

NOTES

1. The classic account is Thomas S. Kuhn, *The Structure of Scientific Revolutions* (Chicago: University of Chicago Press, 1962).
2. Peter J. Bowler, *The Non-Darwinian Revolution: Reinterpreting a Historical Myth* (Baltimore: Johns Hopkins University Press, 1988).
3. These events are recounted in Maria Edgeworth (ed.), *Memoirs of Richard Lovell Edgeworth* (2 volumes, London: R. Hunter, 1820), volume 2, pp. 208–28.
4. *Ibid.*, pp. 137–8. For more details of Edgeworth's achievements see Professor Jim Dooge's essay in this volume.
5. See Desmond King-Hele, *Erasmus Darwin* (New York: Scribners, 1963), pp. 19, 25–6 and 42.
6. These letters are reprinted in Edgeworth's *Memoirs*, volume 2, pp. 130–4, 155–8 and 173–8. See Erasmus Darwin, *Zoonomia: or the Laws of Organic Life* (2 volumes, London, 1794–96, reprinted New York: AMS Press, 1974).
7. See Roy Porter, 'Erasmus Darwin: Doctor of Evolution?', in James R.

Moore (ed.), *History, Humanity and Evolution: Essays for John C. Greene* (Cambridge: Cambridge University Press, 1989), pp. 39–69.

8. William Paley, *Natural Theology*, in *The Works of William Paley* (5 volumes, London: F.C. and J. Rivington *et al.*, 1819), volume 4, see pp. 337–45 (although Darwin is not identified by name).
9. Adrian Desmond, *The Politics of Evolution: Morphology, Medicine and Reform in Radical London* (University of Chiocago Press, 1989).
10. See James Secord's introduction to his edition of Robert Chambers, *Vestiges of the Natural History of Creation and Other Evolutionary Writings* (University of Chicago Press, 1994).

William James MacNeven: Chemist and United Irishman

William J. Davis

In the 17th century a Scientific Revolution occurred in Europe which laid the foundations for our present-day science. This revolution inspired the dazzling programme for the extension of knowledge in the 18th century that we know as the Enlightenment, that pursuit of understanding of everything in the universe, including the behaviour of man himself, by the application of rational empirical scientific procedures. Early on, the new philosophy hardly touched Ireland, where, by the mid-1600s the land was being laid waste by Cromwell and others. The Irish-born Robert Boyle made his considerable contribution to the Scientific Revolution, but he did so outside Ireland; the main Irish contribution came from the rents paid by his Irish tenants which allowed Boyle the leisure to carry on his intellectual pursuits in London and Oxford. In a letter to Samuel Hartlib, the great scientific intelligenser, who acted as a conduit for much of the scientific information then being generated in Europe, Boyle makes plain the situation in 17th century Ireland.

> [This] is a barbarous country, where chemical spirits are so misunderstood and chemical instruments so unprocurable that it is hard to have any Hermetic thoughts in it.[1]

Under Hartlib's urging Boyle participated briefly with the intellectual group of like-minded men in Ireland, mostly of English origin, but with a few Protestant Irishmen such as William Molyneux and Allan Madden. This group was to form the Dublin Philosophical Society,[2] which although short-lived — it existed for a mere 25 years or so — was to sow the seeds, which were to influence developments of science in Ireland in the 18th century. The principal reason for its decline was the political unrest in Ireland. Following the battle of the Boyne, the last decades of the 17th century and the first decade of the new century saw the introduction of the Penal Laws that were to deny education to the bulk of Ireland's indigenous Roman

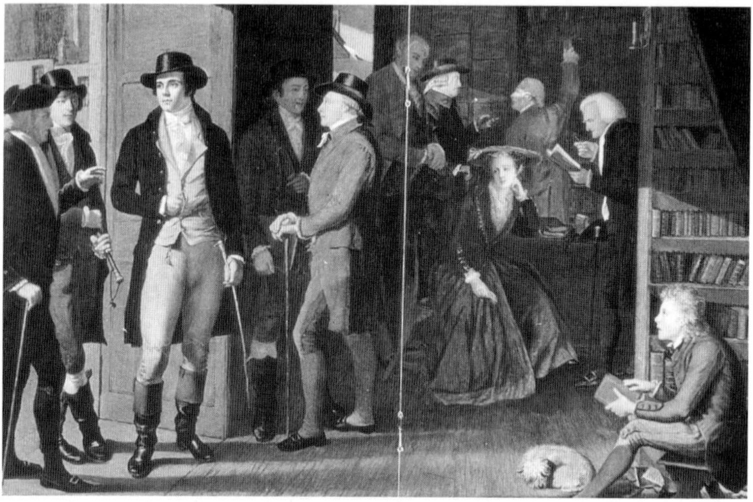

Figure 1. Sibbald's Circulating Library *by W.B. Johnstone (1786)*

Figure 2. The United Irish Patriots of 1798. W.J. MacNeven is in the centre seated at the front.

Catholic population. It is instructive to compare the situation in Ireland with that in Scotland where immense contributions to the new way of thinking were made. A painting (Figure 1) by W.B. Johnstone entitled *Sibbald's Circulating Library* (1786) depicts some of Edinburgh's leading lights in this intellectual movement. It includes two doctors, one a professor of rhetoric, the other Professor of Natural Philosophy, a publisher, two artists, Alexander Naismith and David Allan, Lord Monboddo, a judge and anthropologist, James Bruce the explorer of Africa, and the poet Robert Burns. Eventually the Enlightenment was a powerful enough movement to touch Ireland. One of its manifestations was the foundation of the Dublin Society in 1731 (later to be the Royal Dublin Society). Its first members were of Anglo-Irish stock, second and third generation Irish born: they were all members of the ruling ascendancy. Many were extensive landowners and it is significant that for the first sixty years of the Society the activities were centred narrowly on land improvement. Also significant is that the members were, with few exceptions, of the established Church in Ireland and not Roman Catholic, the religion of the majority of the Irish population. The opportunities for Catholic involvement in official or intellectual activities in Ireland were essentially non-existent for the vast majority of the Irish population, because of the Penal Laws introduced after the accession of King William III. This created a social climate very different from the Scottish one, and directed liberal thinkers along a very different path. Another picture, produced in 1898 for the centenary celebrations of the rising by the Society of the United Irishmen, (Figure 2) and entitled *The United Irish Patriots of 1798* purports to show the principle figures of the rebellion. They include S. Neilson, M. Dwyer, J. Sheares, W. Corbet, A. O'Connor, A.H. Rowan, W. Jackson, W.J. MacNeven, M. Teeling, R. Emmet, H. Sheares, T.W. Tone, J.N.B. Tandy, T.A. Emmet, J. Hope, H.J. McCracken, T. Russell, and Lord E. Fitzgerald. Seated centre stage is the subject of this paper, William James MacNeven (Figures 2 and 3).[3] Unlike the majority of the Directorate of the United Irishmen, MacNeven was a Roman Catholic. He himself had his early career shaped by the Penal Laws. Born on 21st March 1763 at Ballynahowna, near Aughrim, County Galway of a land-owning family, he was sent at the age of 11 to live in Prague with his paternal uncle, O'Kelly MacNeven, who was Physician to the Empress Marie Theresa. After a sound classical education William James MacNeven studied medicine at Prague and Vienna, where he obtained his M.D. degree in 1783.[4] In the following year he started practice as a doctor in Dublin. He was a bright and accomplished man, speaking French and German, a little Italian, as well as Irish and English. His years in Central Europe,

Figure 3. William James MacNeven.

one of the most active areas for the mining of metals[5] must have extended his interest in science outside medicine and one of his first publications was a translation of a pamphlet in the German language on the use of a Mine-auger, published in 1788 (Figure 4).[6] This essay was dedicated to the Dublin Society. In the Preface he refers to 'the excellent work of Mr. Kirwan ... already out of print' and goes on:

> Mountains that consist of stones of the argillaceous genus, and of the fixth compound species of the siliceous genus (gneiss) are the principles seat of the metallic substances.

After the quotation[7] he goes on:

> Hence it appears, it is only the torch of experiment that can illuminate their recesses; their outward appearances are uncertain lights, and for this reason have often involved people in the most ruinous expense; but at a trifling one, the Mine-auger brings the mineral from its concealment, enables us to explore our precious earths and increases together private property with public welfare.

Going on to compare the achievements of the Germans with

E S S A Y

ON THE

CONSTRUCTION AND USE

OF A

MINE-AUGER.

TO WHICH ARE ADDED,

The Defcription of an Earth-Borer,

FOR

COUNTRY GENTLEMEN AND FARMERS;

AND AN IMPROVEMENT ON THE

WELL-BORER OF MERSENNUS.

ILLUSTRATED WITH COPPER-PLATES.

TRANSLATED FROM THE GERMAN OF MR. GEISS,

BY *WILLIAM JAMES MNEVEN*, M. D.

Member of the Royal-College of Phyficians of London.

LONDON:

Printed for W. RICHARDSON, Royal-Exchange.

M,DCC,LXXXVIII.

Figure 4. Title page of MacNeven's pamphlet on the mine-auger (London, 1788)

those of the Irish he not only gives a contemporary view of the state of science in Ireland at the end of the 18th century but also his aspirations for its development.

> There is, however, this in our favour, that beginning at so advanced a period of experimental philosophy we have their numerous acquisitions as lights to guide and spare us mismanagement and errors, otherwise unavoidable in the infancy and first improvements of every art. With such advantages it is presumed we will soon regain our lost ground, especially when assisted by the judicious and patriotic exertions of the Dublin Society, together with the learned labours of our Academy ... it must be highly pleasing to see future Rinmanns and Bergmans[8] arise among us, by cultivating our mines, and a numerous people flourish. Dublin No. 118, Stephen's Green, June 30, 1788.

In about 1790, shortly after this paper was published MacNeven became involved in politics. He came to the United Irishmen through the 'Catholic Committee', the General Committee of the Catholics of Ireland. This body attempted to ameliorate the position of the Roman Catholic population, as regards their status and to bring about the repeal of the Penal Laws. Its strategy was to go over the heads of the local politicians and make direct approaches by petition to London. Some of the wealthier members, such as Lord Kenmare, favoured a petition which was very much a sycophantic address stressing their allegiance to Britain. MacNeven took an active part in the debates of the Committee, and his attitude was much stronger than Kenmare's. He wished to make demands, e.g. for the removal of taxes, such as quarterage, levied by Dublin Corporation on Catholics only. It was his advocacy while on a deputation to London that led to the so-called 40-shilling freeholders being given the right to vote at elections. It was only after the failure of the Catholic Committee to make any further real headway, that he turned to the Society of the United Irishmen; he was sworn in as a member of the secret society in December 1796, a mere two years before the Rebellion. Older than many of his fellow conspirators, his sound negotiating skills were recognised and he quickly became a member of the Leinster Directory. While his vocation as a doctor led him to be a dove as regards an armed rising by Irish civilians (unlike the aristocratic firebrands Arthur O'Connor and Lord Edward Fitzgerald), he favoured as legitimate an invasion by professional French soldiers. He acted as a spokesman for the United Irishmen with the French Government asking for financial and military help. These activities became known to the British authorities.

On the 12th March 1798, the Irish Administration made a pre-emptive strike against the United Irishmen, by arresting many of the Directorate, while they attended a meeting in Oliver Bond's house. MacNeven was arrested in his Dublin rooms, where he was breakfasting with Lord Edward Fitzgerald. The latter was let go because no warrant had been issued for his arrest. This lapse on the part of the authorities was eventually to lead to Fitzgerald's wounding when he was later apprehended, and his subsequent death.

The prisoners, including 16 leading revolutionaries, were housed in Kilmainham Jail, Newgate and the Bridewell. The Administration, after imposing martial law at the end of March, arrested a large number of the rebels in both Dublin and outside. In mid-July Dublin Castle proceeded with treason trials in the civil courts. The first of these, of Henry and John Sheares, was short and not very sweet. In 17 minutes the jury found them guilty and within 24 hours they were hanged. The convictions of McCann, Byrne, and Oliver Bond followed. McCann was executed on the 24th July. After his execution the gentleman prisoners signed a round robin proposing that, without implicating individuals, they would agree to give full details of the United Irishmen movement in Ireland, and of its links with foreigners. In return they should be allowed to emigrate. The hangings had certainly concentrated their minds. As MacNeven put it, 'the best service" they could perform was 'to save the country from the cold-blooded slaughter of its best, its bravest, and its most enlightened defenders". This attitude was in line, as one might expect, with his humanitarian viewpoint as a physician. But it is noteworthy that the Catholic members of the Directory, e.g. MacNeven and Richard McCormick, perhaps more aware of the awful retribution that could be visited by a vengeful Government, had always sided with the Protestant doves, such as Thomas Addis Emmet, and tended not to espouse violence. The firebrands, such as Arthur O'Connor and Lord Edward Fitzgerald, whose families had never suffered the rigour of the Penal Laws represented the hawks of the United Irishmen. After some dithering on the part of the Government, a delay during which the unlucky Byrne was led to the scaffold, and following more specific proposals from the prisoners, a pact (the 'Treaty of Newgate') was signed between the parties and the executions were stopped.The 40-page document eventually produced under the names of Arthur O'Connor, Thomas Addis Emmet, and William James MacNeven was frank in its confession of conspiracy to establish a socialist republic, if necessary by force of arms. One point forcefully made in the document was that the activities of the United Irishmen was not a Roman Catholic conspiracy; they thought this idea laughable, as the Directory were almost exclusively Protestant and free thinking. MacNeven, when later asked by the Secret Committee of the House of Lords and Commons inquiring into the rebellion for his views on the role of the Catholic church in Ireland, replied 'For my part, I would as soon establish a Mahometon [Mohammedan] as the Popish religion".[9]

The prisoners were still in Dublin gaols in January 1799; the Government had hoped to transport them to America, but President John Adams, had declined to take them on the grounds that they were dangerous men to have in the 24-year old Republic.[10] So they were packed off to languish in Fort George in Inverness, Scotland. When, in 1802, the Treaty of Amiens brought a brief respite to the war between France and Great Britain, they were allowed to banish themselves to the land which had been one of the inspirations for their rebellion. Doubtless to recuperate from his long incarceration MacNeven went on a walking tour of Switzerland, and subsequently wrote an account of his travels in a book entitled *A Ramble though Swisserland.*[11] It is an interesting account of his journey, and reveals his wide interests, in agriculture, landscape, military tactics, and the methods of government. His recent treatment by the English obviously still rankled for he writes:

> At the ship, I saw a number of English Carriages before the door, and as I always found an advantage in not stopping at the inns which the English frequented, I went on to the crown, where I found attendance and accommodation excellent. Another gentleman and I dined on a cold dinner of fine baked trout, roast chickens and an excellent ham, veal & salad, dessert of fruit and biscuits, and a mass and a half of wine — two English bottles, for according to the exchange rate of the time, not more than 1s 9d. a piece. Bed and breakfast also cost 1s. 9d.

He goes on to describe the falls at 'Laussene' and compares them unfavourably with the Powerscourt waterfall:

> [A traveller] ... after heavy rains, the fall of Powerscourt in the county of Wicklow, in Ireland, has seen perhaps a finer object.

The French had recently overrun Switzerland and he visited a number of the sites of battles. In an extensive appendix to the book he records the recent history of the country and comments on the 'mediation' by Napoleon in Switzerland. His scepticism on powerful politicians is revealed when he states

> ... let the rulers of nations learn, if it be possible for them to be instructed, that henceforth oppressive governments will not have the support of the people.

MacNeven became a Surgeon-Captain in the Irish Brigade of the French Army, where he served for two years. He tried to interest Napoleon in a further invasion of Ireland, but by 1805 he appears to have abandoned hope of further direct involvement in revolutionary matters. For seven years his professional life had been disrupted by his involvement with the United Irishmen. He was now 42 years old.

In June 1805 he took ship from Bordeaux for the United States of America, arriving paradoxically on the 4th of July, the day of celebration for that other country whose liberty had given inspiration to the United Irishmen. He was welcomed by his friend and former fellow prisoner at Fort George, Thomas Addis Emmet, who had established a legal practice in New York a year previously. Emmet had already established himself as a successful lawyer in New York, and he must have helped MacNeven to become integrated into New York society. Soon he was established in what was to become a successful medical practice. MacNeven married Jane Margaret Riker Tom, daughter of Samuel Riker and widow of John Tom of New York, on 15th June 1810. They had several children: James Joseph who died aged 19 years; William Hugh, who became a doctor; Samuel Riker; and Jane Maryanne, who wrote the account of her father for Madden's *The United Irishmen*.[12] A step-daughter, Anna Tom, was to marry Thomas Addis Emmet (b. 1797), the third son of his friend Thomas Addis Emmet.

His career in America was a most successful one, both as a medical doctor and as a chemist. In June of the following year (1807) he was admitted *ad eundem* as a M.D. in Columbia College, and he was appointed associate physician under David Hosack,[13] botanist and doctor, at the New York Almshouse. MacNeven was appointed in 1808 Professor of Obstetrics in the College of Physicians and Surgeons. His interests were still centred round chemistry, especially the practical aspects of the subject. It must be remembered that medical doctors at this time often had to prepare their own medicines; *materia medica* was one of the subjects taught to students, and MacNeven also held the chair of this subject for some time. Subsequently, following a reorganisation of the Faculty of the College of Physicians and Surgeons in 1811 (brought about by the intrigues of Hosack, the Professor of Botany, who appears to have been a real mover and shaker) MacNeven was assigned to the chair of Chemistry, a subject much more to his liking than those he was teaching. Under his supervision a laboratory was built that enabled students to perform experiments rather than just to witness them as demonstrations. This was the first proper chemical laboratory in New York. He was obviously regarded as the expert in practical chemistry in America as is shown by some correspondence with Reuben Haynes, secretary of the Academy of Natural Sciences of Philadelphia. A letter from MacNeven in 1815 reveals his obvious abilities in laboratory design and management, a subject that must then have been quite new.[14]

[Page 1] New York May the 14th 1815
Mr. Reuben Haynes

Dear Sir,

Dr Hosack showed me a few lines from you expressive of a wish to have some account from me of my laboratory. It is with great pleasure I hasten to comply, and that I shall always be happy at every opportunity of showing respect & attention to your wishes.

The first consideration is to adopt your laboratory to its object, or the processes that are to be performed in it. The laboratory of an apothecary, or for preparing Chemical medicines on a larger scale, or for the application of Chemistry to the arts, or for electrochemical processes, or metallurgic processes, or for lecturing to an indiscriminate class, are all different; and if one were constructed to embrace each object, it is probable a great portion of it would be seldom used; unless indeed [Page 2] the labours were carried on jointly by a few cooperators, which will perhaps be your case. For the arrangement of furnaces I can not refer you to any thing better than what I have already given in the 4th Vol. of the American Medical and philosophical register, and with which from experience I am now perfectly satisfied. I would only add on this subject the very great convenience of setting up in some part of the laboratory a Black's furnace through which you may pass a tube. Indeed if care be taken to feed it copiously with air from the outside of the building, it may be made to answer the purpose of every other furnace for small experiments. A small apartment should be alloted exclusively to the furnace, with Water. An other apartment h[av]ing a well lighted should be appropriated to the nicer part of your apparatus & the processes to be performed by them. This would admit, however, of a sand heat. A small cool closet convenient to the first wd. be very advantageous for the [Page 3] crystallisation of salts. The lecture room would require an open funnel & a descending screen at the back of the Professor's Chair, & according to the Nature of the audience part of apparatus might be safely deposited or not.

I am happy to learn that the Academy of Natural Sciences of Philadelphia is proceeding so prosperously. My best wishes shall always accompany it, & I feel with becoming gratitude the honor it has done me of electing me a corresponding member. For this favour have the goodness to communicate to the Academy my respectful thanks, & to believe me your obliged & obedt.

Humle. Sert.,

W J MacNeven

The laboratory notebooks and lecture notes of some of his students are extant[15] and show his gift of oratory and organisation. His lectures were illustrated with many carefully planned and executed practical experiments, such as the precipitation of silver chloride when hydrochloric acid is added to silver nitrate, and the precipitation of differing coloured sulphides by the action of hydrogen sulphide on antimony, arsenic and lead. These are experiments familiar to present-day students of chemistry.

MacNeven never neglected the importance of chemical theory. The new ideas of Dalton's Atomic Theory[16] were emerging just as MacNeven was arriving in New York. He was instrumental in promulgating the new chemical ideas to American students. In 1819

EXPOSITION

OF THE

ATOMIC THEORY

OF

CHYMISTRY;

AND THE

DOCTRINE OF DEFINITE PROPORTIONS.

BY

WILLIAM JAMES MACNEVEN, M.D.

PROFESSOR OF CHYMISTRY AND MATERIA MEDICA IN THE COLLEGE OF PHYSICIANS AND SURGEONS OF THE UNIVERSITY OF THE STATE OF NEW-YORK.

WITH AN APPENDIX OF CHYMICAL EXERCISES,

BY THE PUPILS OF THE LABORATORY, &c.

NEW-YORK:

PRINTED BY GRATTAN AND BANKS,

Figure 5. Title page of MacNeven's book on Atomic Theory (New York, 1819)

he published his *Exposition of the Atomic Theory of Chemistry and the Doctrine of Definite Proportions* (Figure 5).[17] In the Preface which is addressed to the Students of the University of the State of New York he states

> Examples have not been wanting in the laboratory of the college to show that nothing so efficiently accelerates the progress of the student or gives so much precision to his acquirements, as a speedy initiation into the principles and details of the new doctrine. The University of which you have the honour of being members is distinguished for the respect and favour with which it regards chymical studies; while, in our particular College, they are indispensable, being made an essential part of the medical classes.

Under a section entitled EXPOSITION &c MacNeven writes:

> It has been known, ever since chymists acquired the art of analysing bodies, that substances always enter into chymical combination in determinate proportions. Nitrate of potash, under all circumstances in all situations, consists invariably of 54 parts nitric acid and 46 potash, per cent. Common salt of 46 muriatic acid and 54 soda; .... carbonate of lime is always found to consist of 43.2 carbonic acid and 56.8 lime. Sulphate of barytes of 34.5 sulphuric acid and 65.5 barytes

He goes on to describe the failures to bring about transmutation of one element into another

> ... alchymists ... eternally foiled in this chimerical attempt.

Writing about the combining properties of substances he refers to what

> ... older chymists ... called saturation ... Modern chymists have given this law the name of definite proportion, meaning thereby to express a fact which they have traced more extensively and correctly than their predecessors

He defines the Law of Definite Proportions as

> The quantity of one substance entering into chymical union is limited therefore by that of the other. ... This is the foundation of Dr. Wollaston's[18] sliding rule of chymical equivalents. The fortunate thought occurred to Mr. Dalton 'that those proportional numbers represented the respective weights of the atoms of the combining bodies.'

Before listing many examples of the usefulness of the theory he makes an interesting and sagacious comment on the Irish chemist, William Higgins' claim for priority in proposing the Atomic Theory

> Mr Higgins, professor of chymistry in Dublin, in a work published by him in 1789, made a near approach to the atomic theory. But his own experiments were too few, and there were not data enough furnished by others to supply sufficient materials for the construction of the atomic theory. During the twenty one years that elapsed between the publication of the comparative view of Mr. Higgins and the elements of chymical philosophy by Dalton, with more industry, the former might, perhaps, have secured for himself the reputation for which he is now so solicitous; but, if liberal and candid, he will confess that celebrity without labour is neither legitimate nor attainable in the walks of science, and that the fame of a discovery belongs rightly to him, who is not only the most sedulous, but also the first to promulgate and apply it. Without being the discoverer of the atomic theory, no name in Britain has done so much for it than Dr. Thompson; and by Berzelius and Gay Lus[s]ac its principles have been carried as much beyond where Dalton left them, as he

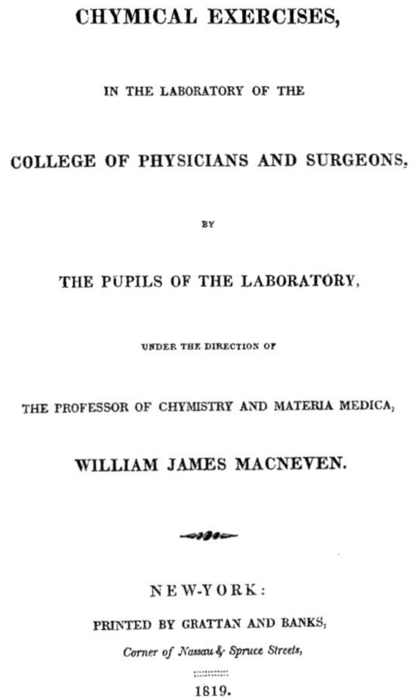
CHYMICAL EXERCISES,

IN THE LABORATORY OF THE

COLLEGE OF PHYSICIANS AND SURGEONS,

BY

THE PUPILS OF THE LABORATORY,

UNDER THE DIRECTION OF

THE PROFESSOR OF CHYMISTRY AND MATERIA MEDICA,

WILLIAM JAMES MACNEVEN.

NEW-YORK:

PRINTED BY GRATTAN AND BANKS,

Corner of Nassau & Spruce Streets,

1819.

Figure 6. Title page of MacNeven's book on chemical analysis (New York, 1819)

exceeded the scanty suggestions furnished by Mr. Higgins. It is now every where received, and constantly improving.

MacNeven's prowess as an early exponent of the Atomic Theory was recognised by his being asked to write about it in Thomas Thompson's *Annals of Philosophy* in 1820.

As a teacher MacNeven appears to have been outstanding. As an appendix to his Atomic Theory he produced a series of *Chymical Exercises* (Figure 6), being mainly chemical analyses of minerals, carried out by his students, including some by J.P. Emmet, the second son of his friend Thomas. He was later to hold the Chair of Chemistry at the University of Virginia.[19] It was MacNeven's intention to produce a full-scale textbook of chemistry for his students. Instead of producing an original text, in 1824 he published an emended version of W.T. Brande's *Manual of Chemistry* (Figure 7),[20]

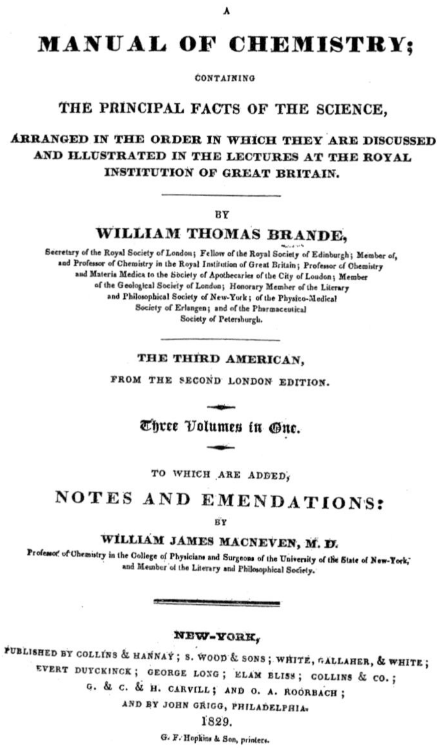

A

MANUAL OF CHEMISTRY;

CONTAINING

THE PRINCIPAL FACTS OF THE SCIENCE,

ARRANGED IN THE ORDER IN WHICH THEY ARE DISCUSSED AND ILLUSTRATED IN THE LECTURES AT THE ROYAL INSTITUTION OF GREAT BRITAIN.

BY

WILLIAM THOMAS BRANDE,

Secretary of the Royal Society of London; Fellow of the Royal Society of Edinburgh; Member of, and Professor of Chemistry in the Royal Institution of Great Britain; Professor of Chemistry and Materia Medica to the Society of Apothecaries of the City of London; Member of the Geological Society of London; Honorary Member of the Literary and Philosophical Society of New-York; of the Physico-Medical Society of Erlangen; and of the Pharmaceutical Society of Petersburgh.

THE THIRD AMERICAN,

FROM THE SECOND LONDON EDITION.

Three Volumes in One.

TO WHICH ARE ADDED,

NOTES AND EMENDATIONS:

BY

WILLIAM JAMES MACNEVEN, M. D.

Professor of Chemistry in the College of Physicians and Surgeons of the University of the State of New-York, and Member of the Literary and Philosophical Society.

NEW-YORK,

PUBLISHED BY COLLINS & HANNAY; S. WOOD & SONS; WHITE, GALLAHER, & WHITE; EVERT DUYCKINCK; GEORGE LONG; ELAM BLISS; COLLINS & CO.; G. & C. & H. CARVILL; AND O. A. ROORBACH; AND BY JOHN GRIGG, PHILADELPHIA.

1829.

G. F. Hopkins & Son, printers.

Figure 7. Title page of MacNeven's emended version of Brande's Manual of Chemistry *(New York, 1829)*

that was very successful, and ran into several editions. It became the standard text-book of American students of chemistry.

William James MacNeven was to become an active and prominent citizen of New York. With his friend Emmet he founded the New York Athenæum, at which he often lectured on the utility of chemistry. He played an important role in putting the case for Catholic Emancipation and never lost interest in Irish affairs. He looked after the welfare of many Irish immigrants who landed in New York and even published a welfare guide for their use.

Neither did his rebellious spirit leave him. In the 1820s a dispute arose between some of the professors and the Regents (the administrators) of the New York College of Physicians and Surgeons. It was a complex affair, involving alleged overcharging of fees, interference in examinations, and the appointment of younger staff

Figure 8. Rutgers Medical College, Duane Street, New York city.

to the Faculty.[21] The four major participants, MacNeven, Samuel L. Mitchill, Valentine Mott, and Hosack decided to resign from the College and found their own medical school. This they did under the patronage of Rutgers College in New Jersey. The new college opened in November 1826 in Duane Street in New York city (Figure 8). It was an immediate success, indicating the high regard for the medical teaching of its professors including MacNeven. Its success was short-lived, however, for the Regents won in the end by seeking its closure by legal means in 1830.

After the closure of the Duane Street school MacNeven, who was 67 years old, decided not to return to teaching but to devote himself to his private practice. Perhaps he was tired of revolutions that ended in failure. He was to live for another eleven years, dying on the 12th July 1841, as a result of an accident when his carriage was overturned. He was buried in the Riker burial plot at Bowery Bay, Long Island. His friends and colleagues erected a 40-foot tall monument to his memory in St. Paul's Protestant churchyard on Lower Broadway (Figure 9) close to one he had raised to his friend T.A. Emmet. The citation states:

WILLIAM JAMES MACNEVEN, M. D.
BORN IN IRELAND, MARCH 21TH A.D. 1763.
DIED IN NEW YORK, JULY 12TH A.D. 1841

WHO IN THE CAUSE OF HIS NATIVE LAND
SACRIFICED THE BRIGHT PROSPECTS OF HIS YOUTH
AND PASSED YEARS IN POVERTY AND EXILE;
TILL IN AMERICA HE FOUND A COUNTRY
WHICH HE LOVED AS TRULY AS HE DID
THE LAND OF HIS BIRTH

TO THE SERVICE OF THIS COUNTRY WHICH
HAD RECEIVED HIM AS A SON
HE DEVOTED HIS HIGH SCIENTIFIC ACHIEVEMENTS
WITH EMINENT ABILITY

DR MACNEVEN

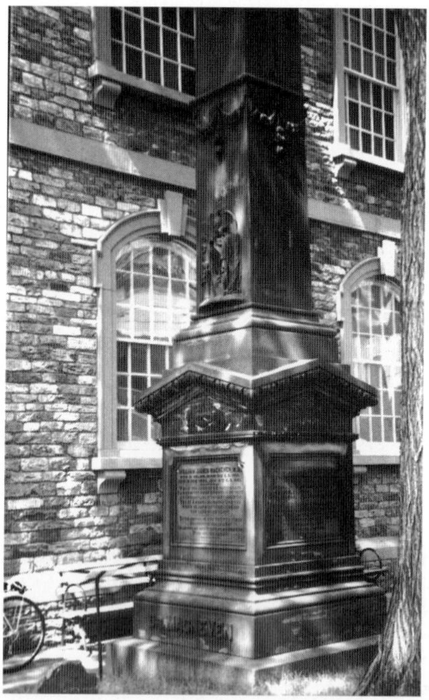

Figure 9. Base of 40-foot pillar commemorating MacNeven in St Paul's churchyard in Lower Broadway, New York city.

NOTES

1. Letter from Robert Boyle to Samuel Hartlib (1651) quoted in *Dictionary of National Biography.*
2. Karl Theodore Hoppen, *The Common Scientist in the Seventeenth Century: a study of the Dublin Philosophical Society, 1683-1708* (London: Routledge & Kegan Paul, 1970).
3. D. O Raghallaigh, 'William James MacNeven: 1763-1841', *Studies*, 30 (1941): pp. 247-59; Richard Robert Madden, *The United Irishmen, their lives and times* (2nd series) (7 volumes, London: J. Madden, 1843), volume 2, pp. 214-5; H.A. Kelly and W.L. Burrage, *Dictionary of American Medical Biography* (Longwood Press, 1979), pp. 796-7; R.M. Jones (ed.), *Family History, from the Writings of Jane MacNeven Purdy and others* 10. (New York: Long Island Historical Society, Brooklyn); M.H. Saffron, 'The Influence of Dublin Medicine', *Proceedings 23rd International Congress of History of Medicine*, Volume 1 (1974): pp. 841-5; Byron Stookey, 'William James MacNeven (1763 - 1841) Versatile Professor', *Bulletin of the New York Academy of Medicine* 41(10) (1965): pp. 1037-51; J.P. Murray, 'W.J. MacNeven: Father of American Chemistry', *Irish Medical Journal*, 79 (9) (1986); D. Reilly, 'An Irish-American Chemist, William James MacNeven, 1763-1841', *Chymia* 2 (1949): pp. 17-26.
4. Davis Coakley and Zdenek Kalvach, 'Doctors in exile: William MacNeven O'Kelly (1713-1787) and William James MacNeven (1763-1841)', in Helga Robinson-Hammerstein, *Migrating Scholars: Lines of Contact between Ireland and Bohemia* (Dublin: Trinity College Library, 1998), pp. 81-115.
5. Bohemia and Saxony were the prime mining areas of Europe since the 15th century. In the 18th century the German mineralogist A.G. Werner put forward a set of ideas about the formation of the Earth's crust from fluid sources (the Neptunist theory), which although later shown to be incorrect, helped launch the new science of geology.
6. William James MacNeven, *Essay on the Construction and Use of a Mine-auger. To which are added, the description of an Earth-borer, for Country Gentlemen and Farmers; and in Improvement on the Well-borer of Mersennus* (London: W. Richardson, 1788). Translated from the German edition of Mr Geiss. [Marin Mersenne was a 17th century French mathematician and Minim Friar]
7. From Richard Kirwan's *Elements of Mineralogy* (London: P. Elmsly, 1874).
8. Sven Rinmann and Torbern Bergman were Swedish chemists much concerned with mining and minerals.
9. William James MacNeven and Thomas Addis Emmet, *Pieces of Irish History, illustrative of the condition of the Catholics of Ireland, of the Origin and Progress of the Political System of the United Irishmen and of their transactions with the Anglo-Irish Government* (New York: Bernard Dornin, 1807) pp. 174-93. Also in Castlereagh Correspondence L. 353-372.
10. See J.P. Murray, note 3.
11. William James MacNeven, *A Ramble through Swisserland in the Summer and Autumn of 1802* (Dublin: Stockdale, 1803).

12. See B. Stookey, note 3.
13. Charles Coulston Gillispie (ed.) *Dictionary of Scientific Biography,* Volume 4, (1970), p. 521.
14. Letter from W.J. MacNeven dated 14 May 1815 to Reuben Haines, Philadelphia in the R. Haines Collection, Haverford College Libraries, Connecticut [Box 2, Manuscript M. 3.]
15. C.A. Browne, 'Lecture and Laboratory Notebooks of Three Early Irish-American Refugee Chemists', *Journal of Chemical Education,* (April 1941): pp. 153–8. Two small notebooks, which belonged to Thomas Arnold Brayton, a medical student, in which are recorded some of MacNeven's 1823/4 course at the New York College of Physicians and Surgeons, are held in the New York Public Library [MS 70–1713].
16. John Dalton, *A New System of Chemical Philosophy* (London: Bickerstaff, 1808).
17. William James MacNeven, *Exposition of the Atomic Theory of Chemistry and the Doctrine of Definite Proportions* (New York: Gallatin and Brooks, 1819). Also reprinted in *Annales de Chimie* in France.
18. Wilfred Hyde Wollaston was an English chemist who discovered the element Palladium. He developed a slide rule to calculate the combining ratios of chemicals.
19. See article by D.T. Burns in this volume (p. 25).
20. William James MacNeven, *A Manual of Chemistry by William Thomas Brande. First American edition from the second London edition, to which are added notes and emendations by William James MacNeven, M.D., Professor of Chemistry in the College of Physicians and Surgeons of the University of the State of New York* (1821). Another edition appeared in 1829. William Thomas Brande FRS (1809) was Superintendent of chemical operations in Apothecaries Hall, London in 1812. He succeeded Humphry Davy as Professor of Chemistry at the Royal Institution (1813–54), with Michael Faraday as 2nd Professor from 1827. His textbook ran to five editions.
21. A complete account of the dispute and the founding of the Rutgers Medical School is given by Christine C. Robbins, 'David Hosack. Citizen of New York', *Memoirs of the American Philosophical Society,* 6 (1964): pp. 1–246.

John Patten Emmet (1796-1842): foundation Professor of Chemistry in the University of Virginia

D. Thorburn Burns

INTRODUCTION

To claim John Patten Emmet as an Irish Chemist in 1798 is not as such true, he was but 2 years old in 1798 and at best a putative chemist. However, it is true to say that his life and career in chemistry were both shaped by the events of 1798.

Fortunately, thanks to his son, Thomas Addis Emmet, there is available a good account of John Patten Emmet's life and times.[1] This account includes much detail from contemporaneous correspondence within the Patten family, and on which all later accounts are based.[2, 3]

EARLY LIFE AND EDUCATION

John Patten Emmet (Figure 1) was born in Dublin on the 8th of April 1796. He was the second son of Thomas Addis Emmet, lawyer and previously a physician, one of the leaders of the United Irishmen, and Jane, daughter of the Rev. John Patten, a Presbyterian minister of Clonmel.[4] He was also the nephew of the great Irish patriot Robert Emmet.

His father was imprisoned for his part in the 1798 uprising, first in Dublin, and later in Fort George, Scotland. After release in 1802 Thomas Addis Emmet lived in Brussels and then Paris, prior to emigration to the United States in 1804. During his father's period at Fort George, after his mother was allowed to join his father, John remained in Dublin and lived with his grandparents. In March 1805 in the company of his younger sister and brother and a family friend he emigrated to join his parents in New York. Just prior to leaving Ireland he suffered a severe attack of smallpox, then measles followed by whooping cough. After this his constitution remained impaired for the rest of his life.

Figure 1. John Patten Emmet (1796-1842)
[From Thomas Addis Emmet 1898 [1]]

Figure 2. Mary Byrd Tucker Emmet
[From Thomas Addis Emmet 1898 [1]]

Young Emmet was educated by Richard W. Thompson, a graduate of Trinity College Dublin, at a noted classical school located at Flatbush, Long Island in the State of New York.

In 1814 Emmet became a cadet at the United States Military Academy, West Point, and stayed until 1817. The young cadet was an excellent draughtsman/artist, and student and had such mathematical talent that he was appointed acting assistant professor in the subject. For reasons of health he did not graduate from West Point. He then spent a year abroad, mainly in Italy studying languages, music and art.

Upon return home early in 1819 he began the study of medicine in the College of Physicians and Surgeons under the direction of Dr William J. MacNeven, the Professor of Chemistry, a close family friend. He graduated in 1822, having written a thesis entitled 'The Chemistry of Animated Matter'. He moved to Charleston to practice his profession and as he was not fully occupied gave a series of public lectures on science which brought him to the attention of Thomas Jefferson, Rector of the University of Virginia.

PROFESSOR OF CHEMISTRY AND LIFE AT THE UNIVERSITY OF VIRGINIA

In 1825 John Patten Emmet was appointed to be Professor of the School of National History; as from 1827 he was Professor of Chemistry and Materia Medica.

The letter of the initial appointment is interesting as it gives details of the emoluments and conditions, both of which were excellent. It reads

> MONTICELLO, March 6th, 1825
>
> Dear Sir:
>
> The board of visitors of the University of Virginia, at their last meeting of the 4th inst., proceeding to the appointment of a Professor for the school of Natural History in that Institution, unanimously nominated you to that chair. Under the general term Natural History, they comprehend Zoology, Botany, Mineralogy, Chemistry, and Geology; that of Chemistry however being considered as the branch most eminently distinctive of the School.
>
> Its emoluments are fifteen hundred dollars a year fixed salary, tuition fees from those of your school from twenty-five to fifty dollars each according to circumstances, and an excellent house and convenient garden grounds for your residence. The tenure of the Professorship is under a Board of Visitors, seven in number, two thirds of whom, say five

out of seven, can alone remove a Professor. It is therefore a freehold in fact. But one vacation is admitted, to wit from the 15th of December to the last day of January, and lectures are expected every other day during term. As you are probably a stranger to this establishment, I have thought it right to state to you these particulars. The Institution opens tomorrow, so that in the hope you will accede to our wishes, we shall request your attention as early as possible; and in the meantime ask an answer which may place us on a certainty. Accept the assurance of my great esteem and respect.

THOS. JEFFERSON

In June 1827 Dr Emmet became engaged to be married to Miss Mary Boyd Farley Tucker (Figure 2) a native of Bermuda, who was then, with her mother, visiting her Uncle George Tucker, the first Professor of Moral Philosophy in the University of Virginia. They married on the 21st July the same year and had unalloyed happiness for a period of fifteen years of married life. The records show he was an excellent father and well liked by his children, who he amused with his drawings.

Thomas Addis Emmet recorded his mother's description of how she took possession of the residence of the Professor of Natural

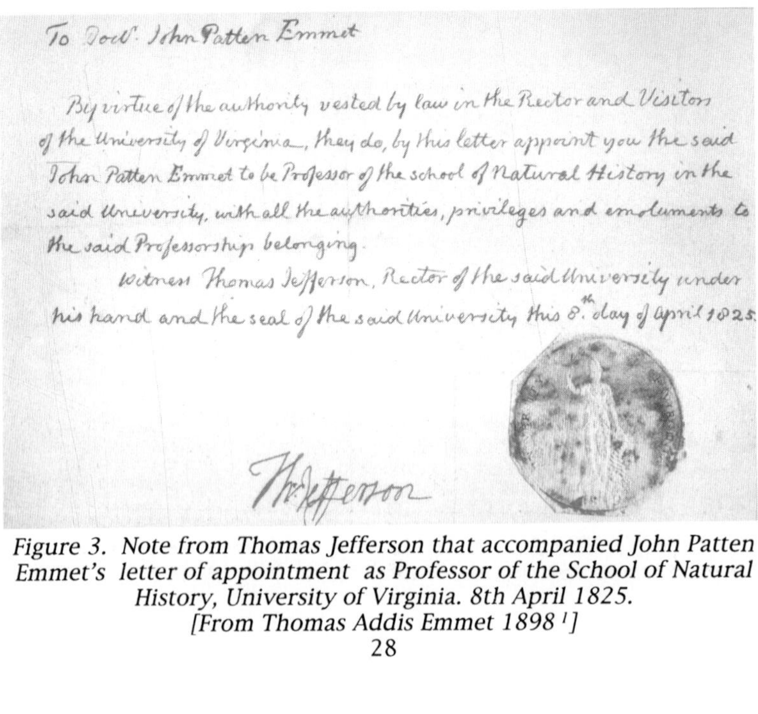

To Doct. John Patten Emmet

By virtue of the authority vested by law in the Rector and Visitors of the University of Virginia, they do, by this letter appoint you the said John Patten Emmet to be Professor of the school of Natural History in the said University, with all the authorities, privileges and emoluments to the said Professorships belonging.

Witness Thomas Jefferson, Rector of the said University under his hand and the seal of the said University this 8th day of April 1825.

Th: Jefferson

Figure 3. Note from Thomas Jefferson that accompanied John Patten Emmet's letter of appointment as Professor of the School of Natural History, University of Virginia. 8th April 1825. [From Thomas Addis Emmet 1898 [1]]

History. The largest and best room was devoted to a collection of snakes which were retarded in their movements by having the floor waxed - the snakes were banished in a summary manner as if St Patrick had exercised his power to that end!

In 1834 Emmet bought land adjoining the University and built a house 'Morea' (Figure 4). Here he conducted agricultural experiments. In January 1842 failing health obliged him to visit Florida, and in May, somewhat recovered, he set sail for New York. The ship was damaged in a hurricane and drifted for 38 days prior to being rescued and taken to New York. His health was so damaged that he died on the 15th August 1842.

EMMET'S CONTRIBUTIONS TO SCIENCE AND EDUCATION

During Emmet's time at the University he conducted courses and research in chemistry and physical topics. His published chemical studies[5-10] concerned:

(a) A test for arsenic based upon precipitation of a double salt with

Figure 4. Morea, the residence of John Patten Emmet, near the University of Virginia. [From Thomas Addis Emmet 1898 [1]]

Figure 5. An original sketch by John Patten Emmet, from White Street and Broadway looking towards Canal Street, New York.

potassium iodide and subsequent liberation of iodine.[5]

(b) Study of the hydrolysis of ammonium salts and of reactions of molten ammonium salts such as ammonium nitrate.[6]

(c) Presence of bromine and of iodine in Kenawha waters.[7]

(d) Isolation of the bitter principle of *Liriodendron tulipifera* (American poplar).[8]

(e) Experiments on the setting of gypsum.[9]

(f) Formation and properties of formic acid.[10]

His published physical papers[11-14] are those on:

(a) Generation of electrical pulses using a magnet and coil-wrapped keeper.[11]

(b) Two papers on thermoelectric effects using a variety of metal couples.[12, 14]

(c) The causes of voltaic currents in coils and revolving plates.[13]

His early work on diffraction was never published.

It has been stated that Emmet made no important contributions to science.[3] This is perhaps a little harsh in that several of his papers were considered of sufficient interest at the time to be republished in other journals than the primary one.[5, 8, 9, 10] Emmet published mainly in *The American Journal of Science and Art* often referred to as Silliman's Journal after its editor Benjamin Silliman. Silliman was of the opinion that Emmet's papers were important overall and that the physical papers showed ingenuity and originality.[15] In 1852 when Dr J. Lawrence Smith was called to the professorship of chemistry at Virginia he informed Benjamin Silliman of 'Emmet's valuable and extended notes of original experiments on light and other subjects not yet published'.[15] It is interesting to note Silliman's connection to Ireland via his membership of the Belfast Natural History Society. Emmet worked hard at teaching, kept his courses up-to-date and with good demonstrations.

At least one set of contemporaneous lecture notes exists,[16] those of Haller Nutt for the session 1833-34. They are not complete and relate mainly to heat, light and electricity. From the notes it is clear that Emmet gave great attention, by mention and by demonstration, to apparatus.

Among the pieces of apparatus referred to by Nutt in his notes of Emmet's lectures are the goniometers of Hauy and of Wollaston; the parabolic reflector; Leslie's apparatus for measuring the absorption of solar radiation by different surfaces; Franklin's device for comparing the absorption of heat by differently coloured bodies; Wollaston's cryophorus; Daniel's hygrometer; the barometer; the Reaumur, Centigrade, and Fahrenheit thermometers; Rutherford's registering thermometer; Leslie's thermometer; Wedgwood's pyrometer; Dufey's electrical machine; the Leyden jar; Franklin's lightning conductor; Galvani's device for showing the effect of electricity on muscular action; Volta's electric pile; Volta's Couronne des tasses or series of electric cells; Cruikshank's trough battery; Wollaston's battery; Hare's calorimeter; Oersted's galvanometer; the pneumatic trough; the gasometer; and various other experimental devices. So far as one can judge from Nutt's record book, Emmet's lectures seem to have been built largely around demonstrations of apparatus.

CONCLUSION

Despite his ill health and short life, John Patten Emmet made, through his experimental research and teaching, a significant contribution to early chemistry in the United States of America.

NOTES

1. Thomas Addis Emmet, *The Emmet Family with some incidents relating to Irish History and a Biographical sketch of Prof. John Patten Emmet M.D. and other Members* (New York: Privately printed, 1898).
2. Howard Atwood Kelly and Walter L. Burrage, *Dictionary of American Medical Biography: Lives of eminent physicians of the United States and Canada, from the earliest times* (Longwood Press, 1979).
3. Clark A. Elliott, *Biographical Dictionary of American Science: the Seventeenth through the Nineteenth Centuries.* (Westport: Greenwood Press, 1979).
4. Alfred John Webb, *A Compendium of Irish Biography: comprising sketches of Distinguished Irishmen, and of Eminent Persons connected with Ireland by office or by their writings.* (Dublin: Gill and Son, 1877).
5. J.P. Emmet, 'The Iodide of Potassium, (Hydriodate of Potassa of the Shops) as a test for Arsenic, with remarks upon the natures and properties of the compound formed', *American Journal of Science and Art*,18 (1830): pp. 58-63; *Royal Institution Journal*, 1 (1831): pp. 173-4.
6. J.P. Emmet, 'Upon the solvent and oxidating powers of Ammoniacal salts, etc.', *American Journal of Science and Art*, 18 (1830): pp. 255-60.
7. J.P. Emmet, 'Bromine and Iodine in Kenawha Waters, Va.', *American Journal of Science and Art*, 18 (1830): p. 260, (1830).
8. J.P. Emmet, 'Liriodendrine, or the Bitter Principle of the Liriodendron Tulipifera or American Poplar', *Journal of Philadelphia College of Pharmacy*, 3 (1832): pp. 5-17; *J. de Pharm.*, 17 (1831): pp. 400-6.
9. John P. Emmet, 'Experiments upon the Solidification of raw Gypsum', *American Journal of Science and Art*, 23 (1833): pp. 209-12; *Edinburgh New Philosophical Journal*, 15 (1833): pp. 69-73; *Dingler Polytecha. J.*, 49 (1833): pp. 447-51; *J. de Pharm.*, 19 (1833): pp. 685-6; *Liebig Annal.*, 7 (1833): pp. 346-9.
10. J.P. Emmet, 'Formic Acid - remarks upon its utility, together with a correction of Dobereiner's Process and views respecting the theory of its formation', *American Journal of Science and Art*, 32 (1837): pp. 140-8; *Bibl. Linir.*, 11 (1837): pp. 172-8.
11. J.P. Emmet, 'An Enquiry into the cause of the Voltaic Currents produced by the action of magnets and electro-dynamic cylinders upon coils and revolving plates', *American Journal of Science and Art*, 26 (1834): pp. 23-44, (1834).
12. J.P. Emmet, 'Notices on Thermo-electricity and Electro-magnetism', *American Journal of Science and Art*, 26 (1834): pp. 311-3.
13. J.P. Emmet, 'A New Mode of developing Magnetic Galvanism, by which may be obtained shocks, vivid sparks and galvannic currents from the Horse-Shoe Magnet', *American Journal of Science and Art*, 24 (1833): pp. 78-86.
14. J.P. Emmet, 'Upon Caloric, as a cause of Galvanic Currents', *American Journal of Science and Art*, 25 (1834): pp. 269-81.
15. B. Silliman, 'American Contributions to Chemistry', *The American Chemist*, 5 (1874): pp. 70-114, 195-209.

16. C.A. Browne, 'Lecture and Laboratory Notebooks of Three Early Irish-American Refugee Chemists, William J. Macneven, John P. Emmet and Thomas Antisell', *Journal of Chemical Education*, 18 (1941): pp. 153-8.

Tumultuous times: geology in Ireland and the debate on the nature of basalt and other rocks of north-east Ireland between 1740 and 1816

Patrick N. Wyse Jackson

INTRODUCTION

This paper focusses on the debate on the nature of rocks on the north-east coast of Ireland in the late 18th and early 19th centuries, and examines the rôle of protagonists on both sides, especially that played by the Reverend William Richardson (1740–1820). It also examines the state of geology in Ireland in the 1790s and it comments on geological activity in the Dublin Society, Trinity College Dublin, and the Royal Irish Academy.

THE STATE OF GEOLOGY IN IRELAND IN THE LATE 1700s[1]

Prior to 1750 little was known of Ireland's geological structure. Only a few papers on Irish geological phenonema had been published — some by members of the short-lived Dublin Philosophical Society (1683-1708)[2] — namely those by Sir Richard Bulkeley (d. 1710), Bishop Samuel Foley (d. 1695) and Thomas Molyneaux (1661-1733) on the Giant's Causeway.[3] Other early geological surveys were carried out under the aegis of the Physico-Historical Society of Ireland which published four county accounts — of Down, Kerry, Waterford, and Cork — the last of which contains a map on which geological strata are indicated.[4] It wasn't until the mid 1700s when mineralogy became both a popular persuit and a focus for serious scientific study that geology as we now know it really developed.[5]

In Ireland a number of factors promoted geology as a serious subject for academic and economic study. Firstly, a good network of roads was available which allowed gentlemen and scientists ease of travel to more remote parts of Ireland.[6] Secondly, a number of learned societies were established: the Dublin Society in 1731 and

Figure 1. View of Benmore (Fair Head) on the north Antrim coast, showing moderately well developed columnar basalt. It is possible that the figures on the rocks in the foreground are the Rev. William Richardson (pointing out geological structure) and Humphrey Davy (seated and sketching). [Original drawing by D. Stewart (possibly the Itinerant Mineralogist employed by the Dublin Society). Engraving reproduced from William Hamilton Drummond 1811 The Giants' Causeway, a poem. *Belfast: Longman, Hurst, Rees, Orme & Browne].*

the Royal Irish Academy in 1785, while in Trinity College Dublin a geological collection was assembled in its museum that itself had been established in 1777. It was within the walls of these institutions that geological topics were both promoted and debated by members. Within time these institutions developed a high reputation such that until the early decades of the 1800s Dublin ranked close to centres of geological excellence such as Freiburg, Edinburgh and London.

Many individuals also became interested in matters geological. The Lord Bishop, William Augustus Hervey (1730–1803) Bishop of Derry and fourth Earl of Bristol,[7] apparently installed steps down to the Giant's Causeway and made trips to the volcanic regions of Italy. Later when bored of geology he assembled a fine art collection and embarked on a massive building programme of country houses. Both his Irish houses were subsequently demolished after his death but his fine Mussendun Temple still stands.[8]

Thomas Russell (1767-1803), the noted United Irishman, was a keen amateur geologist who collected minerals and rocks in the Mourne Mountains, and ascribed to the Neptunian theory for the formation of the geological succession.[9]

The Dublin Society

The Dublin Society was established in 1731 to promote agriculture and sciences in Ireland.[10] It did so by offering premiums to members to carry out research in mineralogy and for proficiency in the subject,[11] by organising county statistical surveys, by the acquisition of geological (principally mineralogical) and other specimens for their Museum, and by acting as a forum for discussion, debate, and study. Richard Kirwan (1733-1812), the Galway-born chemist, mineralogist and geologist of the highest repute, urged through the Society's *Transactions* that a Board of Mines be established. He suggested that it should have responsibility for all Irish mines, but unfortunately his plans never made it past the printed page.[12]

In 1786 Donald Stewart (d. 1811), appointed as Itinerant Mineralogist, studied the geology of various parts of the country and collected specimens for the Society's museum.[13] This Museum contained the large and important collection of 7,331 mineral specimens assembled by the German mineralogist Nathanael Gottfried Leske (1751-1786), which had been purchased in 1792, largely thanks to the efforts of Kirwan, for the huge sum of £1350 (approximately the equivalent of £35,000 today).[14] This remains the most important geological acquisition ever made by an Irish museum. It stimulated considerable geological activity by members who made mineral collections for themselves, many of which ultimately found their way into the Museum or into other institutions.

Between 1801 and 1832 the Dublin Society published the results of county statistical surveys of twenty-three counties, in which some geological information is given. Of particular note are the four volumes on Counties Wicklow (1801) by Robert Fraser (1760-1831), Kilkenny (1802) by William Tighe (1766-1816), Londonderry (1802) by George Vaughan Sampson (1762-1827) and Cork (1810) by Horatio Townsend (1750-1837) which contained geological maps, that were for their time cartographically and geologically highly advanced.[15]

However, it was through their promotion of art that Ireland's most famous geological site became widely known. In 1840 the Society offered a £25 premium for art and one of the entries was from a Susanna Drury, a Dublin artist. She submitted several canvasses of views of the Giant's Causeway in County Antrim. Painted over a period of three months they were excellent representations of the coastline and of the pillars of columnar basalt.[16] Miss Drury was awarded the premium and subsequently her images were engraved in London by François Vivarès. It was these prints, and later versions by others,[17] that were distributed across Europe, and which initiated a steady stream of visitors to the Causeway coast, whose rocks then became the subject of examination and explanation. Visitors included Archdeacon Richard Pococke in 1747,[18] John Wesley in 1778, John Whitehurst in 1783, Abraham Mills in 1787-88,[19] the Frenchman, De Latocnaye in 1797,[20] Humphry Davy in 1806,[21] and the Swiss geologist Jean-François Berger in 1811.[22]

Trinity College Dublin

Geology has for over 200 years interested the Fellows and students of the University of Dublin.[23] The Dublin University Museum was founded in 1777 by Rev. William Hamilton (1755-1797) who became Fellow of the College two years later. Within two decades of its foundation the Museum had acquired some geological material, notably Italian volcanic rocks collected by the Rev. George Graydon (*c.* 1753-1803),[24] and minerals from various sources, including those purchased in Saxony by the Professor of German.[25] The collections did not impress the celebrated Scottish geologist Robert Jameson (1774-1854) when he visited the Museum on 30 June 1797. He remarked: 'Went to the College Museum ... It is an elegant room, with a few trifling fossils, some Indian dresses &c.'.[26]

The College continued to acquire collections which were added to the displays. By 1807 the Museum consisted of a systematic mineralogical collection of 1,089 specimens, a geological collection of rock types, and a geographical collection comprising material from various parts of the world, including Graydon's Vesuvius material and specimens from Ulster (Stokes 1807), and a wooden model of the Giant's Causeway — unfortunately long since lost. In 1817 another visitor Anne Plumptre wrote: 'the Museum is a good room, and contains a tolerable collection of Irish minerals, with some specimens of the basaltic columns from the Giants' Causeway'.[27]

Geology was not taught as a formal subject until 1816 when it was included in the natural history lecturing remit of Whitley Stokes (1763-1845) — a one-time United Irishman. The Chair of Geology was not established until 1843. Nevertheless, various Trinitymen became interested in the subject: men such as the aforementioned William Hamilton, and the Rev. William Richardson, another Fellow,

Figure 2. Rev. William Hamilton [From Hamilton 1822 [36]]

who is an important player in the present tableaux — whose collections were first housed in the University Chemical Laboratory.[28] Another, William Fitton (1780–1861), a Classics scholar and graduate of 1799, became interested in geology as an undergraduate and was unfortunate to be arrested in Dublin in 1798 in possession of his geological hammer — a dangerous weapon it appears![29] The aristocratic Hon. George Knox (1765–1827), the Member of Parliament for the University, was the owner of a considerable mineral collection, which was later to be purchased by the College.

Royal Irish Academy

The Royal Irish Academy was founded in 1785 for the study of Sciences, Polite Literature and Antiquities.[30] It elected its members from the noble, clerical and academic circles of the day and rapidly built up an important library and museum.

The Academy also developed a Museum that contained important archaeological material and some geological specimens and became the official repository for treasure trove found in

Ireland. Many years later the geological collections were incorporated into the collections of the Dublin Society and are now in the National Museum of Ireland.[31]

Geological problems were among those tackled by members such as Kirwan and Richardson and several papers dealing with James Hutton's theory and the nature of the rocks of north-east Antrim were published in the Academy's *Transactions.* This important journal was published from 1798 until 1907 and copies were sent to similar learned institutions in Europe, thus allowing for a rapid diffusion of the authors' geological ideas.

THE DEBATE ON THE NATURE OF BASALT IN NORTH-EAST IRELAND

In Ireland the debate on the origin of basalt and other rocks naturally centred theoretically and petrologically on whether these rocks were sedimentary or igneous in origin, and geographically around the Giant's Causeway in County Antrim. One body of thought, the 'Neptunists' or to use a contemporary term 'watermen', felt that these rocks were precipitates from water — a view held by the German mineralogist Abraham Gottlob Werner (1749-1817) and his students in the Freiberg School of Mines. They regarded lava as being a product of volcanoes, but held that where volcanoes occurred in basaltic regions, they had formed after the basalt had been deposited. Some were convinced that lithologically basalt was quite distinct from lava, and others that lava was produced when volcanic heat and air acted on pre-existing (presumably water-lain) basalt.

The other group, variably known as 'Vulcanists' or 'Plutonists' or 'firemen', according to the rock types in which they were interested, argued that the rocks were once heated and were the products of volcanoes or other igneous mechanisms.[32] The Vulcanist theory was proposed by the Frenchman Nicolas Desmarest (1725-1815), who, having observed the extinct volcanoes in central France, advocated in 1768 that basalt was the product of volcanic activity.[33] He was quickly followed in this opinion by others including Sir William Hamilton (1730-1803) and John Strange (1732-1799) who were familiar with the volvanic regions in Italy. In 1788 James Hutton (1726-1797), the Scottish geologist, published his paper 'Theory of the Earth' which proposed, amongst other things, that some igneous rocks were plutonic in origin, and argued that these were implaced from a hot magmatic source from below, and that large-scale reworking of older rock successions had taken place in the past.[34]

Firemen -v- Watermen in the Irish context

Nicolas Desmarest never viewed the Giant's Causeway, but on seeing Vivarès' engravings of Susanna Drury's paintings was the first geologist to declare that the basalts that formed the region were volcanic in origin.

In 1786 John Whitehurst suggested that the columnar basalts of the Giant's Causeway formed as the molten basalt cooled.[35] In the same year the Rev. William Hamilton, who had visited the region, published an important and influential memoir entitled *Letters concerning the Northern Coast of the County of Antrim* (Dublin, 1786) which advanced Desmarest's theory of the volcanic origin of the Causeway rocks.[36] He accurately described the geology of the coast. He was one of the founders of the Royal Irish Academy in 1785 and in 1790 he became rector and a local magistrate of Clondevaddog, an isolated Donegal parish. He was murdered in 1797 after local unrest; this agitation culminated in the widespread but unsuccessful uprising of 1798. Hamilton left a widow and nine children who were granted monies by Parliament.

Others, including John Wesley, regarded the basaltic successions in north-east Ireland as being the product of some volcanic fire.

Perhaps the most influential member of the Dublin geological fraternity at the time was Richard Kirwan, President of the Academy from 1799 until 1812.[37] His book *Elements of Mineralogy* (1784, 2nd edition 1794) was well-received by his peers and made him one of the most influential mineralogists of his day. In later life Kirwan's eccentricities were notable; unfortunately these, have clouded the views of historians who have emphasised them rather than concentrating on his not insignificant scientific abilities and achievements. Although many of his ideas, including some geological and those on phologiston were incorrect, Kirwan's ability to argue through the printed page forced others to look at these ideas and counterargue. Such dialogue allowed for new ideas to emerge.

Kirwan was outraged at Hutton's ideas and went into print defending the Neptunian theory in a number of papers from 1793 and in his *Geological Essays,* published in 1799.[38] In doing so he spurred Hutton into writing his book *The Theory of the Earth, with proofs and illustrations,* which eventually appeared in 3 volumes.[39] Eventually much of what Hutton, and indeed Desmarest, wrote was accepted by the global geological community, but not without a fight from the Neptunists.

Some of Kirwan's objections, which are somewhat difficult to glean from the contorted logic and language of his papers, can be summarised briefly: he said that basalt rested on various rock types

Figure 3. Engraving (c. 1809, J. Sidebotham, Dublin), showing Rev. William Richardson scything grass. In the foreground are copies of his published papers. The signpost points to Clonfeacle and Armagh. Richardson published a number of papers in which he encouraged the sowing of fiorin grass, which he felt would reduce Ireland's poverty. [Reproduced with the permission of Dr E.C. Nelson[53]].

including coal, limestone, gneiss, and granite in which no evidence of the effect of heat could be seen. It is now known that gneiss is one of the highest grade metamorphic rocks which is produced by the effect of enormous stress and high temperature. He noted that many basalts contain zeolites, minerals rich in water and calcite, which on account of their chemistry could not have been heated.[40] He failed to realise that the zeolites were deposited subsequent to the formation of the basalt.

He also objected to the findings of Sir James Hall (1761-1832) who had taken basalt, melted it and recrystallised it, in order to show the similarity of the artificial with the real[41] Here was good proof that basalt was igneous in origin. The recrystallised basalt contained minute air bubbles which Kirwan stated were characteristic of basalt deposited in water.

Kirwan also discussed the effects of the Noacean Flood on the landscape and claimed that Galway Bay was the site of a granite mountain that had been swept away during the Flood.[42] Given that the Galway Granite is exposed along the northern margin of the bay, such an explanation is understandable. What Kirwan failed to understand or accept was that slow long-term erosion, and not the floodwaters, was responsible for the removal of material from what is now Galway Bay.

While Kirwan was loud in his rejection of Hutton and Desmarest, his disciple the Reverend William Richardson, was louder, and perhaps with good reason, for he actually spent some of the year living at Portrush, on the Antrim coast. Richardson, a Fellow of Trinity College of 1766, took Holy Orders, and eventually resigned in September 1783 on taking up the living of the parish of Clonfeacle in County Tyrone. Through his membership of the Royal Irish Academy he would have become familiar with the controversy surrounding the Giant's Causeway. During his annual sojourns at Portrush Richardson examined the ground for himself, and he published a steady stream of papers between 1802 and 1812 which outlined his views on the geology of the region.[43] He clearly was upset at others pontificating on the geology of an area which they had either not visited at all or had examined for just a short time. He believed he had the authority, as a long-term resident, to make his views known.[44] While he claimed not to subscribe to any theory or belong to any faction,[45] he clearly if unwittingly held to the Neptunist theory for the origin of the basalt. He rejected the alternative theories on a number of grounds:

1. There was no evidence of a volcanic mountain or cone in Antrim.
2. Plants are found developed on lava flows but not in between the basalt layers.
3. The constituents of basalt and lava were different.
4. The layers of basalt were horizontal and regular in thickness.
5. The physical appearence of basalt and lava were different.
6. The contact relationships of basalt and lava were different.
7. The basalt was divided into regular masses while lava is found as a irregular mass.
8. The basalt of Portrush contained fossil marine shells.

Given the limited exposure Richardson had to other volcanic regions his objections to the Vulcanist and Plutonist theories are not surprising. The ammonite-bearing stone was discovered by Richardson in about 1799 and was proof to him and many others of the sedimentary origin of basalt. He also located specimens of basalt in a quarry at Ballylagan in which cavities were infilled with 'fresh water'. To him this was stronger evidence of the sedimentary origin of basalt than was the presence of fossils.[46] Richardson's material was sent to Trinity College Dublin, Kirwan and other geologists, and later to Plutonists in Edinburgh.

From the perspective of the 1798 uprising Richardson is interesting as it was he, together with a local landowner Thomas Knox, who first suggested in June 1796, the formation of armed units of local tenants (later referred to as the yeomanry corps) to defend the property and lives of the landed classes against the upsurgent activists, whose agitation culminated in the 1798 rebellion.[47] One of these activists was Thomas Russell who was personally known to Richardson and had indeed dined with him in the Glebe house in Clonfeacle in 1792.[48] Later Richardson turned his pen to agricultural subjects and wrote extensively on the merits of growing fiorin grass. He died in September 1820, continuing to the last to hold onto his outdated geological viewpoint.

RESOLUTION AND CONCLUSIONS

Specimens of the fossil-bearing 'basalt' were sent to Edinburgh in 1801, where they were examined by Sir James Hall, Lord Webb Seymour and John Playfair (1748–1819) who identified the specimens as being sedimentary cherty hornstone.[49] The rock is now known to be Lower Jurassic Lias mudstone which contains specimens of the ammonite *Paltechioceras*.[50] The Portrush rock had been baked by hot basaltic material (actually dolerite of the now named Portrush Sill) close by, and which had resulted in its similar appearance to that basalt. Their conclusions were confirmed by Jean-François Berger, and by the Reverends William Conybeare (1787–1857) and William Buckland (1784–1856) in 1816.[51] This piece of evidence was critical, and swung the scales in the favour of the Vulcanists and Plutonists, whose view held that basalt was derived from a molten source.

Additional evidence on the true nature and origin of basalt came from later observations made in Italy and elsewhere of active and extinct volcanoes where the presence of columnar basalt was seen in unequivocal lava flows. In terms of the Antrim basalts it is now known that they were vented from fissures and that a volcanic cone was never present in north-east Ireland; that plants are preserved in the interbasaltic horizons; and that the regularity of the columns is due to slow-cooling of a ponded viscous lava.

Today the Giant's Causeway continues to attract and excite geologists and the general public alike,[52] and its importance was confirmed by its designation some twenty years ago as a UNESCO

World Heritage Site. While there had to be a winner and a loser in the debate on its origins and composition, the Causeway Coast in north-east County Antrim remains one of the most important sites in the history of geology. It is a matter of some pride, to Irish historians of science and geology, that much of the sabre-rattling pertaining to the debate on the nature of basalt during the late 1700s and early 1800s should have taken place partially in Ireland.

ACKNOWLEDGEMENTS

I am grateful for the long-term advice and encouragement of Gordon L. Herries Davies and Colain MacArthur. I thank Dr E.C. Nelson for permission to reproduce the print of Richardson in his possession.

NOTES

1. Gordon Herries Davies' valuable bibliographic essay covers the period before 1812: 'Geology in Ireland before 1812: a bibliographic outline', *The Western Naturalist*, 7 (1978): pp. 84-99.
2. Karl Theodore Hoppen, *The Common Scientist in the Seventeenth Century: a study of the Dublin Philosophical Society, 1683-1708* (London: Routledge & Kegan Paul, 1970).
3. Richard Bulkeley, 'Part of a letter from Sir R.S. S.R.S. to Dr. Lister, concening the Giant's Causeway in the County of Atrim in Ireland. Old Bawn Apr. 24, 1693', *Philosophical Transactions of the Royal Society of London* 17 (1693): pp. 708-10; Samuel Foley, 'An account of the Giants Causway in the north of Ireland', *Philosophical Transactions of the Royal Society of London*, 18 (1694): pp. 170-5; Thomas Molyneux, 'Some notes upon the foregoing account of the Giants Causway, serving to further illustrate the same', *Philosophical Transactions of the Royal Society of London*, 18 (1694): pp. 175-82.
4. Gordon L. Herries Davies, 'The making of Irish Geography IV: The Physico-Historical Society of Ireland 1744-1752', *Irish Geography*, 12 (1979): pp. 92-8; Gordon L. Herries Davies, *Sheets of Many Colours — the Mapping of Ireland's Rocks 1750-1890* (Dublin: Royal Dublin Society, 1983).
5. Rachel Laudan, *From Mineralogy to Geology: the Foundations of a Science, 1650-1830* (University of Chicago Press, 1987).
6. George Taylor and Andrew Skinner, *Taylor and Skinner's Maps of the Roads of Ireland, Surveyed 1777* (London: Published for the authors, 1778). A revised edition appeared in 1783 [this second edition was republished in facsimile, with a new introduction by J.H. Andrews, in 1969 by Irish University Press, Shannon]; Peter O'Keeffe, *The Development of Ireland's Road Network* (Dublin: Institution of Engineers of Ireland, 1973); Peter O'Keeffe, *Alexander Taylor's Roadworks in Ireland, 1780-1827* (Institute of Asphalt Technology, Irish Branch, 1996).
7. William Shakespear Childe-Pemberton, *The Earl Bishop: the life of*

William Hervey, Bishop of Derry, Earl of Bristol (London: Hurst & Blackett, 1925).
8. T.G.F. Paterson, 'The Edifying Bishop: "Bishop's Folly"', *Quarterly Bulletin of the Irish Georgian Society*, 9(3 & 4) (1966): pp. 67–83.
9. Christopher John Woods, *Journals and Memoirs of Thomas Russell* (Dublin: Irish Academic Press, 1991).
10. Henry Fitz-Patrick Berry, *A History of the Royal Dublin Society* (London: 1915); James Meenan and Desmond Clarke (eds), *RDS: the Royal Dublin Society 1731–1981* (Dublin: Royal Dublin Society, 1981).
11. See Anonymous, 'Premiums offered by the Dublin Society. For answering at public examination in mineralogy', *Transactions of the Dublin Society*, 2 (1801): pp. 24–6; Anonymous, 'Premiums offered by the Dublin Society. For answering at public examination in mineralogy', *Transactions of the Dublin Society*, 3 (1803): pp. 39–41; Anonymous, 'London Society Premiums. Premiums offered by the Society for discoveries and improvements in chemistry, dyeing and mineralogy', *Transactions of the Dublin Society*, 3 (1803): pp. 93–103.
12. Richard Kirwan, 'A plan for the introduction and establishment of the most advantageous management of mines in the Kingdom of Ireland', *Transactions of the Dublin Society*, 1 (1801): pp. 277–84; Richard Kirwan, 'A plan for the introduction and establishment of the most advantageous management of mines in the Kingdom of Ireland', *Transactions of the Dublin Society*, 2 (1802): pp. 245–51.
13. Donald Stewart, 'The report of Donald Stewart, Itinerant Mineralogist to the Dublin Society', *Transactions of the Dublin Society*, 1 (1799): pp. 1–142; see also C.W.P. MacArthur, 'Mineralogical and geological travellers in Donegal 1787–1812', *Donegal Annual*, 39 (1987): pp. 39–57.
14. N.T. Monaghan and E. Vaccari, 'I minerali di Giovanni Arduino nella collezione geo-mineralogica di Nathanael Gottfried Leske: verifica di un caso di communicazione scientifica nell'Europa del tardo settecento', *Geologica Romana*, 29 (1993): pp. 547–65.
15. Gordon L. Herries Davies, note 4, second reference.
16. M. Anglesea and J. Preston, ''A philosophical landscape': Susanna Drury and the Giant's Causeway', *Art History*, 3 (1980): pp. 252–73.
17. These later prints included an excellent series of three views, of the east and west prospects of the Giant's Causeway, and of the Isle of Staffa which illustrated a work by the French geologist Déodat Guy Silvain Tancrède Gratet de Dolomieu (1750–1801).
18. Pococke, who became successively Bishop of Ossory in 1756 and Meath in 1765, penned three notes on the Giant's Causeway to the Royal Society which were Neptunistic in content: R. Pococke, 'An account of the Giants Causeway in Ireland', *Philosophical Transactions*, 45 (1748): pp. 124–7; R. Pococke, 'A father account of the Giants Causeway in Ireland, by the Rev. Richard Pococke LL.D. Archdeacon of Dublin, and F.R.S.', *Philosophical Transactions*, 48 (1748): pp. 226–37; R. Pococke, 'A letter upon the same subject from the Rev. Richard Pococke LL.D. Archdeacon of Dublin, to the Rev. Tho. Birch D.D. secr. R.S.', *Philosophical Transactions*, 48 (1753): p. 238. He suggested that vertical pillars (of basalt) formed of precipitated muds became compressed by overburden pressure into the polygonal shapes typically seen at the Giant's Causeway.

19. See Gordon L. Herries Davies, note 1.
20. De Latocnaye, *A Frenchman's Walk through Ireland 1796-7* (Belfast: McCaw, Stevenson & Orr, 1917). [1984 edition published by Blackstaff Press, Belfast].
21. Robert Siegfried and Robert H. Dott, jr., *Humphry Davy on Geology: the 1805 Lectures for the General Audience* (Madison: University of Wisconsin Press, 1980), p. xxxiv. Some of Davy's sketches made on his trip in 1805 are reproduced. It is probable that Richardson acted as his guide during part or all of his sojourn in Antrim.
22. C.W.P. MacArthur, 'Dr Jean-François Berger of Geneva (1779-1833): from the Travelling Fund to the Wollaston Donation', *Archives of Natural History*, 17 (1990): pp. 97-119.
23. For histories of geology in Trinity College Dublin see Patrick N. Wyse Jackson (ed.), *In Marble Halls: Geology in Trinity College, Dublin* (Dublin: Department of Geology, Trinity College Dublin, 1994); and G.L. Herries Davies, 'Geology' in Charles H. Holland, (ed.), *Trinity College Dublin and the Idea of a University* (Dublin: Trinity College Dublin Press, 1991), pp. 294-311.
24. For further details of Graydon's travels in Italy and his collections see: E. Vaccari and P.N. Wyse Jackson, 'The fossil fishes of Bolca and the travels in Italy of the Irish cleric George Graydon in 1791', *Museologia Scientifica*, 12 (1995): pp. 57-81; and Patrick N. Wyse Jackson and Ezio Vaccari, *The Reverend George Graydon (c. 1753-1803): Cleric and Geological Traveller* (Dublin: Royal Irish Academy, 1997).
25. P.N. Wyse Jackson, 'The Geological Collections of Trinity College, Dublin', *The Geological Curator*, 5 (1992): pp. 263-74.
26. J.M. Sweet, 'Robert Jameson's Irish Journal, 1797', *Annals of Science*, 23 (1967): pp. 97-126.
27. Anne Plumptre, *Narrative of a Residence in Ireland during the Summer of 1814, and that of 1815* (London: H. Colburn, 1817). Portions of the basaltic columns are still stored in the present Geological Museum in the University. It is unlikely that they are those referred to.
28. Richardson's collections cannot now be recognised in the present Geological Museum in the College. His important suite of basalts and ammonite-bearing sediments numbering 22 specimens in total, is described in a manuscript catalogue of Robert Perceval's mineral collection: 'Doctor Richardson's explanation of his studio of specimens from the Giant's Causeway, Portrush and adjoining coasts.' This important manuscript is now in the Department of Geology, Trinity College Dublin.
29. P.N. Wyse Jackson, 'William Henry Fitton (1780-1861) and the Wollaston Medal of 1852', *Geoscientist*, 10 (1998): p. 9. Fitton's contribution to Dublin geology was his editing and contribution to the late Rev. Walter Stephen's memoir published in 1812: Walter Stephens and William Fitton, *Notes on the Mineralogy of Part of the Vicinity of Dublin* (London: William Phillips, 1812).
30. Tarlach O Raifeartaigh (ed.), *The Royal Irish Academy: a Bicentennial History 1785-1985* (Dublin: Royal Irish Academy, 1985).
31. N.T. Monaghan, 1992. 'Geology in the National Museum of Ireland', *The Geological Curator*, 5 (1992): pp. 75-282.
32. G.L. Herries Davies, 'The Neptunian and Plutonic Theories', in David G.

Smith (ed.), *Cambridge Encyclopaedia of Earth Sciences* (Cambridge: Cambridge University Press, 1981).

33. See K. Taylor, 'Nicholas Desmarest and geology in the eighteenth century', in Cecil J. Schneer (ed.) *Towards a History of Geology* (Cambridge, Massachusetts: M.I.T. Press, 1969), pp. 339–356, for an account of Desmarest's work on volcanic theory. Taylor points out that Desmarest's first public airing of his views came in 1765 but was published three years later in the *Encyclopédie* ('Basalte d'Auvergne' *Recueil de Planches, sur les Sciences, les Arts Libéraux, et les Arts Méchaniques avec leur Explication.* Volume 4, Paris: Briasson, David, Le Breton, 1768). However, his theory received widespread attention with the publication of his subsequent monograph on the subject: N. Desmarest, 'Sur l'origine et la nature du basalte á grandes colonnes polygonales, determinés par l'histoire naturelle de cette pierre, observé en Auvergne', *Memoires de l'Academie des Sciences*, 87 (1771): pp. 705–775.
34. J. Hutton, 'The theory of the Earth; or an investigation of the laws observed in the composition, dissolution, and restoration of land upon the globe', *Transactions of the Royal Society of Edinburgh*, 1 (1785): pp. 209–304. The source of the heat was a matter of further debate until the late 18th century. Some argued that the source of heat that produced the igneous rocks was derived from a deep source while others suggested that the source was shallow from combustion of coal and oil etc. See the excellent article by E. den Tex, 'Clinchers of the basalt controversy: empirical and experimental evidence', *Earth Sciences History*, 15 (1996): 37–48, for a full discussion of both the sedimentary versus volcanic question and the depth of heat question.
35. John Whitehurst, *An Inquiry into the Original State and Formation of the Earth; deduced from Facts and the Laws of Nature* ([2nd edition], London: Printed for W. Bent, 1786).
36. William Hamilton, *Letters concerning the Northern Coast of the County of Antrim; respecting the Antiquities, Manners and Customs of that Country, together with the Natural History of the Basaltes, and its attendant Fossils in the Northern Counties of Ireland; illustrated by an accurate Map, and Engravings; in Two Parts.* (London: G. Robinson and P. Elmsly, 1786) [Further editions in 1790 and 1822 — the latter contains a memoir of the author].
37. Biographical accounts of Richard Kirwan can be found in the Dictionary of National Biography as well as in the following: D. T. Burns, 'Richard Kirwan: chemist, mineralogist, geologist and meteorologist 1733–1812', in R.C. Mollan, W. Davis, and B. Finucane (eds), *Some People and Places in Irish Science and Technology* (Dublin: Royal Irish Academy, 1985), pp. 14–15; F.E. Dixon, 'Richard Kirwan: the Dublin philosopher', *Dublin Historical Record*, 24 (1971): pp. 53–64; M. Donovan, 'Biographical account of the late Richard Kirwan, Esq., President of the Royal Irish Academy, &c.', *Proceedings of the Royal Irish Academy*, 4, Appendix 8, (1850): pp. lxxxi–cxviii; P.J. McLaughlin, 'Richard Kirwan: 1733–1812', *Studies*, 28 (1939): pp. 461–474, 593–605; 29 (1940): pp. 71–83, 281–300; E.L. Scott, 'R. Kirwan', in Charles Coulston Gillispie (ed.) *Dictionary of Scientific Biography*, Volume 7, (New York: C. Scribner's Sons, 1973), pp. 387–390; and E.L. Scott, *The Life and Work of Richard Kirwan,*

(London: University of London Ph.D., 1979).

38. R. Kirwan, 'Examination of supposed igneous origin of stony substances', *Transactions of the Royal Irish Academy*, 5 (1793): pp. 51–81; Richard Kirwan, *Geological Essays* (London: D. Bremner, 1799).
39. James Hutton, *The Theory of the Earth, with Proofs and Illustrations* (Edinburgh, 1795).
40. Kirwan's geological ideas are amply dealt with by Charles Coulston Gillispie, *Genesis and Geology* (Harvard Historical Studies, 1951).
41. J. Hall, 'Experiments on Whinstone and Lava', *Transactions of the Royal Society of Edinburgh*, 5 (1805): pp. 43–75. For an illustrated account of these experiments see V.A. Eyles, 'Sir James Hall, Bt. (1761–1832)', *Endeavour*, 20 (1961): pp. 210–16. See also P.J. Wyllie, 'Hutton and Hall on theory and experiments: the view after 2 centuries', *Episodes*, 21 (1998): pp. 3–10 for a modern petrologist's viewpoint.
42. Gordon L. Herries Davies, *The Earth in Decay* (New York: Elsevier, 1969).
43. Richardson's papers included: W. Richardson, 'On basalts', *Philosophical Magazine*, 13 (1802): pp. 129–36; W. Richardson, 'Account of the whynn dykes in the neighbourhood of the Giant's Causeway, Ballycastle, and Belfast', *Transactions of the Royal Irish Academy*, 9 (1803): pp. 21–43; W. Richardson, 'Inquiry into the consisting of Dr. Hutton's theory of the Earth with the arrangements of the strata and other phenomena on the basaltic coast of Antrim', *Transactions of the Royal Irish Academy*, 9 (1803): pp. 429–87; W. Richardson, 'Remarks on the basalts of the coast of Antrim', *Transactions of the Royal Society of Edinburgh*, 5 (1805): pp. 15–20; W. Richardson, 'On the Volcanic Theory. Part 3. Arguments against the volcanic origin of basalt, derived from its arrangement in the County of Antrim, and from other facts observed in that country', *Transactions of the Royal Irish Academy*, 10 (1806): pp. 87–107; W. Richardson, 'A letter on the alterations that have taken place in the structure of rocks on the surface of the basaltic country in the counties of Derry and Antrim', *Transactions of the Royal Society*, 98 (1808): pp. 187–222; W. Richardson, 'On the basaltic productions, &c, of the county of Antrim', Appendix II, in J. Dubordieu, *Statistical Survey of the County of Antrim with Observations on the Means of Improvement, drawn up by the Order and for the Consideration of the Dublin Society* (Dublin: Dublin Society, 1812).
44. He castigates John Whitehurst for not dismounting from his horse to examine the Portrush rock closely: see W. Richardson, 'On the Volcanic Theory. Part 3. Examination of the facts and opinions, given by different advocates for the volcanic origin of basalt, who followed Mr. Desmarest; to wit, Mr. Faujas de St. Fond, Mr. Dolomieu, Mr. Whitehurst, Bishop Troll, Abbe Spalanzani, and Dr. Hamilton', *Transactions of the Royal Irish Academy*, 10 (1806): pp. 63-86: p. 76. See also W. Richardson, 'Inquiry into the consisting of Dr. Hutton's theory of the Earth with the arrangements of the strata and other phenomena on the basaltic coast of Antrim', *Transactions of the Royal Irish Academy*, 9 (1803): pp. 429–87.
45. See Richardson, note 43, in John Dubourdieu, *Statistical Survey of the County of Antrim with Observations on the Means of Improvement, drawn up by the Order and for the Consideration of the Dublin Society*

(Dublin: Dublin Society, 1812).

46. The water-bearing basalt was presented to Trinity College Dublin: specimen number 4 in Richardson's studio of specimens. (See note 28 for details of the manuscript catalogue).
47. William Richardson, *History of the Origin of the Irish Yeomanry* (Dublin, 1801); A.T.Q. Stewart, *The Summer Soldiers: the 1798 Rebellion in Antrim and Down* (Belfast: Blackstaff Press, 1995): pp. 21–2.
48. C.J. Woods, note 9: p. 178.
49. John Playfair, *Illustrations of the Huttonian Theory of the Earth* (Edinburgh, 1802).
50. H.E. Wilson and P.I. Manning, 'Geology of the Causeway coast', 2 volumes. *Memoir of the Geological Survey of Northern Ireland*, Sheet 7 (1978).
51. J.-F. Berger, 'On the geological features of the north-eastern counties of Ireland', *Transactions of the Geological Society of London*, 3 (1816): pp. 121–95; W. Conybeare and W. Buckland, 'Descriptive notes referring to the outline of sections presented by a part of the coasts of Antrim and Derry', *Transactions of the Geological Society of London*, 3 (1816): pp. 196–216.
52. For an excellent guide to the geology of the region see Paul Lyle, *A Geological Excursion Guide to the Causeway coast* (Belfast: Environmental and Heritage Service, 1996).
53. For fuller details of the two versions of this caricature see E.C. Nelson, 'A caricature of an Irish naturalist: Revd Dr William Richardson (1740-1820)', *Archives of Natural History*, 27 (2000): pp. 149-51, and M.D. George, *Catalogue of political and personal satires* (London: British Museum, 1952).

Science, Engineering and the Military in Ireland 1760-1800

James C.I. Dooge

INTRODUCTION

To the question of whether anything interesting was happening in relation to engineering in Ireland in 1798, there are two separate answers. Firstly, engineering as a profession was becoming more firmly based on science and was reaching the end of the transition from military engineering to civil engineering. Secondly, the most difficult challenges facing these civil engineers arose from the new enthusiasm for canal building. Both of these developments are discussed below in order to give some idea of the long-term forces for change operating within engineering at the end of the 18th century.

The greater part of this account is based on the engineering activities in Ireland during the closing decades of the 18th century of five individuals. Three of them — Richard Castle, David Ducart and Richard Lovell Edgeworth — were amateur engineers and two of them — Charles Vallancey and William Chapman — professional engineers. It is noteworthy that none of the five was born in Ireland. An additional section on the native engineers was planned but had to be abandoned because sufficient information for a balanced treatment could not be retrieved from the archival material in the time available. The sections on these five individuals are preceded by a section on the emergence of civil engineering and followed by a section on the role in Ireland of some of the leading consulting engineers of the day from Britain. In the case of Edgeworth and Vallancey we are able to include information on their public activities immediately prior to and during the turbulent year of 1798.

Over half of the references used in the study were from primary sources in the form of manuscripts, printed reports, parliamentary papers, papers published in contemporary journals, and contemporary books. Acknowledgement is due to the assistance of the staff of the National Library, the library of the Royal Irish Academy, the Oireachtas library, the Irish Architectural Archive, the library of the Institution of Engineers of Ireland and to the staff of the Institution of Civil Engineers Archives in London, and of the Civil Engineering Historical Collection in Imperial College London.

THE EMERGENCE OF CIVIL ENGINEERING

Evolution of Engineering as a Profession

Science and engineering are so closely linked today that it is sometimes difficult to realise that this was not always the case. Theodore von Karman, key figure in the theory of turbulence and of flight, characterised the distinction as follows 'Scientists explore what is and Engineers create what never was". Rudyard Kipling has described engineers as the Sons of Martha in a poem of that title and characterised their role in the world as one of making life easier for the sons of Mary. The situation of engineering at the opening of the period under review is well covered in *A History of Technology* by Singer *et al.* Volume 3 (1957) and in particular in the chapter on 'Military Technology' by A.R. Hall and the chapter on 'Canals and River Navigation before 1750' by A.W. Skempton.[1] The first public use of the word civil engineer in contrast to military engineer has been variously dated as 1782, 1768 and 1763. In 1763 John Smeaton and Thomas Yeoman described themselves as 'surveyor and civil engineer' in Mortimer's London street directory. Hence the adoption of the period 1760-1800 in the title of this chapter.

In the first half of the 18th century, French influence was dominant in engineering throughout Europe as a result of the work and the writings of such figures as Vauban (1632-1707) in fortifications, Belidor (1693-1761) in hydraulic structures, Perronet (1708-1794) in bridge building and d'Alembert (1717-1783) in fluid mechanics. The key developments in the evolution of civil engineering as a profession also occurred in France.[2] Professionalisation of military engineering was promoted by the formation of the Corps de Génie in 1676 and the Corps des Ponts et Chaussées in 1716. The need for formal instruction soon became apparent and was met by the formation of the École du Génie at Mézierè in 1749 and the École de Ponts et Chaussées at Paris in 1775. Though the treatment of the topics was elementary at first, the standard gradually improved and notable scientists were among the faculty.[3] This form of engineering education survived the fall of the ancien regime in 1789 and was reinforced by the foundation of the École Polytechnique in 1794 and its continuance since then as the leading Grand École of the French system.

Influence of Continental Ideas

The above measures by the French Establishment of the day were paralleled by an emphasis on science and technology by many of the prominent writers of the Enlightenment. The scope of the famous *Encyclopédie* was influenced by the interest of Diderot in technology and of his main collaborator d'Alembert in science. This factor may well have reinforced the technical interests of Richard Lovell Edgeworth who was greatly interested in the ideas of the Enlightenment.

The combination of an interest in the new thinking and in engineering matters was not confined to landowners like Edgeworth. Another amateur who shared these two disparate interests was Thomas Paine whose *Rights of Man* was the most influential book on political radicalism in Ireland in the years leading up to 1798. Special reading clubs were set up to study it in Belfast and Dublin and those who could not read met to hear it read to them.[4] In a recent political biography of Thomas Paine, the first fifteen pages of the chapter entitled the *Rights of Man* are devoted to Paine's interest in bridges, his lobbying in Paris in 1787 to have his design of a wrought-iron arch bridge (based on the form of a spider's web) adopted by the Academy of Sciences for the site of the Pont de la Concorde, and his efforts in 1788–1790 to have his design adopted in England.[5] In intervals of constructing models of his design and lobbying for its adoption, Paine jotted down notes for what he termed 'my political bridge' i.e. the essay published in 1791 as *The Rights of Man.*

The Continental ideas that shaped the evolution of modern engineering reached Ireland by various routes. Of the amateur engineers treated here, Castle and Ducart were born on the Continent and in their youth saw around them examples of pioneering engineering projects, while Edgeworth encountered Continental thinking through his reading and discussion groups. Vallancey, though born on the Continent, was brought up in England and his connection with key Continental figures was through his translation of their written works in order to supplement his army pay and support his large family. William Chapman spent four years of his youth as captain of a merchant ship sailing between England and the Continent. John Smeaton at 31 years of age undertook a 5-week technical grand tour of the Netherlands in order to improve his competence as a consultant on millworks and has left an account of this journey.[6]

Moves towards professional collegiality in Britain

The first move towards professional collegiality among Civil Engineers in Britain took place when John Smeaton and Thomas Yeoman took the initiative in forming a dining club of Civil Engineers in London in 1771. Initially there were seven members of this Society of Civil Engineers and they met 'once a fortnight at the King's Head Tavern in Holborn from Christmas to the end of the sitting of the Parliament'. The Society soon increased to a membership of about twenty and remained at this figure for a number of years. In 1774, William Jessop, a former pupil of Smeaton, became the first Secretary of the Society and took over responsibility for the minutes from the President, Thomas Yeoman. The Society was reorganised on a more professional basis in 1792 with eight engineering members, eight members associated with engineering, and six honorary members. Known since the 1820s as the Smeatonian Society of Civil Engineers, the Society still exists and meets today, the composition since 1913 being forty-eight engineering members and twelve honorary members.

The original form of the Society was in accordance with the 18th century practice of forming learned dining clubs. In the previous decade the Lunar Society of Birmingham had taken a similar form and provided the forum in which Richard Lovell Edgeworth met the pioneering industrial engineer Matthew Boulton, Josiah Wedgewood and other companions of the scientist and poet Erasmus Darwin. As a result of presenting to the Society for the Encouragement of Arts an account of his design of a phaeton incorporating Erasmus Darwin's idea for avoidance of instability on turning and also an improved method of disengaging unruly horses, Edgeworth had made the acquaintance of Darwin who became a close friend. At about the same time he was introduced by Mr Kier of Birmingham, to another society which met at Young Slaughter's Coffee House in London where he met Sir Joseph Banks, President of the Royal Society, Captain Cook and a number of prominent engineers including John Smeaton and Robert Mylne.

Shortly after the close of the period under review, another society of Civil Engineers was founded by a group of young engineers in London and became a Chartered Institution in 1828.[7] On that occasion, the President-elect Tredgold drafted the definition of Civil Engineering the first sentence of which is still widely quoted to-day: 'Civil Engineering is the art of directing the great sources of power in nature for the use and benefit of man'.

ARCHITECT-ENGINEERS IN IRELAND

From River Navigation to Summit-Level Canals

As mentioned in the introduction, canal building was a major feature of engineering in the 18th Century and the final decade of the 1790s in Britain and Ireland is sometimes referred to as the era of canal mania. Prior to 1760, the main works of inland navigation in these islands were largely concerned with improved river navigation. These works were quite extensive with over 2,000 miles of river improvement.[8] Two Irish canal projects from the 18th century — the Newry Navigation and the Tyrone Navigation — are of particular interest in the history of canal engineering in these islands. The Newry Canal was the first summit level canal in Britain or Ireland and Ducart's Canal serving the Tyrone coalfields was the first canal to use inclines rather than locks between adjacent levels. Both were the responsibility of individuals whose main achievements were in architecture and therefore are good examples of the pre-professional civil engineer.

Richard Castle and the Newry Navigation

Richard Castle who came to Ireland in 1728 was from Hesse-Cassel in Germany, having either been born there or moved there at an early age as a Huguenot refugee. His real name as given in his will was De Richardi but when practising in Ireland he preferred to call himself Richard Castle. He was also known as Cassel, Cassels, Castel or Castles. He had travelled widely in Germany, the Low Countries and France before coming to England. Apparently, Castle came to Ireland in 1728 when aged about 30 at the invitation of Sir Gustavus Hume who was planning a country seat on the shores of Lower Lough Erne. He worked as an assistant architect in the office of Edward Lovett Pearce who was appointed Surveyor General in 1730 and thus responsible for inland navigation under the Act passed in 1729 by the Irish Parliament. Castle wrote a manuscript Essay on Inland Navigation which is in the National Library and was recently reproduced and edited.[9] This appears to be the first available account of canal engineering in English and is well organised and practical. Following Pearce's death in 1733, Castle was in sole charge of the Newry Canal until his replacement in 1736 by Thomas Steers of Liverpool.[10] In 1736 he published an 'Essay towards supplying Dublin with Water' which was severely criticised by another pamphleteer Stokes.[11]

The Newry Canal was proposed to bring coal by water from East

Tyrone to Dublin via Newry and so break the monopoly of English coal on the Irish market. The Newry Canal was the first true summit level canal in these islands and Castle probably built the first stone lock in Ireland.[12] The Newry Canal, which was not completed until 1742, was 18 miles in length with a summit at a level of 78 ft. O.D. near Scarva between Poyntzpass and Terryhoogan. There were nine locks between Newry and the summit and five locks from the summit to its junction with the Upper River Bann just south of Portadown.

An anonymous note in *Anthologia Hibernica* for October 1793 describes Richard Castle as 'a man of the strictest integrity and highly esteemed by the nobility and gentry not only as an artist but as an agreeable companion'.[13] The writer comments that Castle 'sacrificed much to Bacchus, and when in Dublin passed his evenings with Dr Mosse of the hospital, and a few more at a tavern, which they seldom left before three or four in the morning. He was whimsical in some things: he had an aversion to shaving, and was cautious of those he employed'. In regard to his professional work the author notes as follows: 'when the effect of his works was not such as he liked, he frequently pulled them down, and whenever he came to inspect them, he required the attendance of all the artificers who followed him in a long train'. One wonders if his dismissal from the direction of the Newry Canal arose from the disapproval of his leisure pursuits by the project promoters from Newry and Portadown.

Following his replacement on the Newry project, for some reason not stated in the records, Richard Castle returned to Dublin and practised as an architect until his sudden death in 1751. His best known works are Leinster House and the Rotunda in Dublin, Carton House in County Kildare and Powerscourt House in County Wicklow.[14]

David Ducart and the Tyrone Navigation

David Ducart has even more spellings of his name than Richard Castle. In his will made in 1780 he gives his name as Daviso de Arcort but he appears in the Irish sources as Davis Ducart or Ducarte or Ducate or Duchart or Dukart as well as Daviso de Arcort. His origin is unknown, but he was described as having been 'long in the Sardinian service'.[15] Ducart is better known as the architect who designed the Limerick Custom House in 1765 and the Cork Mayoralty House in 1766 and described as the last of the Palladians by the Knight of Glin.[16] Ducart is relevant to this account because of the short length of canal serving the Drumglass collieries in County Tyrone commenced in 1767.

As early as 1709 a petition was presented to the Irish Parliament to connect Newry with the Upper River Bann by cutting a canal and during the following twenty years a number of proposals were made to supplement this by building a canal between the coal pits near Coalisland and the River Blackwater and Lough Neagh.[17] The latter proposal known as the Tyrone Navigation was authorised in 1732 and work commenced. In 1753 a petition was presented to the Irish Parliament for assistance in building a roadway 3 miles from Drumglass to Coalisland to reduce the cost of overland transport and a new phase began which dragged on for many years.[18] In 1767 Parliament agreed to a design by Christopher Myers for a canal from Coalisland to Drumglass but then sought a second opinion and the Myers plan was replaced by one proposed by Ducart. Ducart envisaged a three mile canal from Drumglass to Coalisland consisting of four levels with falls of 70 feet, 65 feet, and 55 feet. between them. This design was noteworthy for the first use in these islands of canal inclines which had been used to transport boats from one level to another in China since 350 B.C. and in Italy since 1350 A.D.[19] In a submission to the Irish House of Commons, dated November 17th 1767 and signed Davis Dukart he wrote: 'the Question of whether a Canal or Inland Navigation without locks is preferable to a Canal with locks is so self-evident in the affirmative, that nothing but the positive Directions of the Honourable Committee, could make me enter into a Discussion on that matter'.[20] A manuscript map by Sloane in the collection of the Royal Irish Academy shows the location of these inclined planes which are referred to as 'hurries'.[21] The scheme was reported on by Jessop and Smeaton in 1773 and never became fully operational.[22] The aqueduct over the River Torrent is still extant.[23]

AN IRISH LANDOWNER AND AMATEUR ENGINEER

Early life in England and Ireland

Richard Lovell Edgeworth was born in Bath in 1744, the descendant of a family that came to Ireland in the 1580s. He has left a lively account of his early life which was published after his death by his daughter Maria.[24] His first tutor at the age of eight was the Reverend Patrick Hughes who had taught Oliver Goldsmith. After some months preliminary tutoring, he was sent to school in Warwick in August 1752. A few years later he was sent to Doctor Norris' School at Drogheda and in 1758 was sent to Mr Hynes School in Longford where he spent two years preparing for entrance to Trinity College Dublin. As a youth he was already interested in things mechanical and in physical sports at which he was a first rate performer. His intellectual interests were stimulated through his having the freedom of the Pakenham Hall library of his relative Lord Longford.

In April 1761 he entered Trinity College but left after 6 months. His own account states 'I passed my time in dissipation of every kind. It was, however, but for six months, the only time in my life that I ever spent in such a disgraceful manner'.[25] Disgusted by his own behaviour, he made a clean break when his father removed him from Dublin to Oxford where he entered Corpus Christi College, Oxford in October 1761. He later kept law terms at the Temple but on the death of his father in 1769 he gave up all thoughts of a career in law. While in London, he developed a design for a turnip-cutter (1766) and for an odometer (1767), receiving a silver medal from the Society for the Encouragement of Arts.[26]

For two decades or so his time was divided between England and Ireland until his final return to Edgeworthstown in 1782 where he spent the next thirty years as a resident landlord. The life in London and Edgeworthstown is reflected not only in his memoirs but also in the novels of his daughter Maria, particularly in *Castle Rackrent* (1800) and *The Absentee* (1812). He influenced his daughter's writing and collaborated with her on the application and adaptation of Rousseau's ideas on education.[27]

Irish Volunteer and Yeoman Captain

When Richard Lovell Edgeworth returned from England in 1782, the Irish Volunteers were well established and the Irish Parliament had an increased influence on Irish affairs. He was a delegate to the Volunteer Convention of November 1783 in Dublin and was one of the aides-de-camp of Lord Charlemont who was Commander-in-Chief of the Volunteer Army. He was an active member of the moderate group led by Lord Charlemont which sought parliamentary reform rather than armed confrontation with the Irish Parliament. When the hopes of reform faded in 1785, he retired from public affairs and concentrated on the running of his estate and his mechanical interests. In the same year he was one of the thirty-eight founding members of the Royal Irish Academy.

His interest in public affairs only revived in 1796 as a consequence of the reaction of the administration to the failed attempt of the French to land an army in Bantry Bay. In the next few years, he became estranged from his fellow landlords because he was critical of the activities of some of the yeoman captains and of some of the newly appointed justices of the peace. He was distrusted because he recruited Catholics as well as Protestants into the Corps of Yeomen which he raised in Longford and because of his refusal as a magistrate to allow the use of party labels in respect of those involved in legal disputes.[28]

In 1798 itself, the summer was relatively quiet in the County of Longford but this all changed with the landing at Killala at the beginning of September. The impact of these events on the Edgeworth family, whose seat was in the line of the French advance, has been described in some detail by his daughter Maria.[29] The family withdrew with the Corps of Yeomanry to the town of Longford and Edgeworth was active in advocating resistance rather than retreat. Edgeworthstown House was spared because one of the local rebel leaders, whose mother had been lent badly needed cash by Edgeworth's housekeeper, turned back a large body of rebels at the gates. When the news of the defeat of the French at Ballinamuck near Granard reached Longford, Richard Lovell Edgeworth was almost lynched by a mob of loyalists who had been convinced by a demagogue that he was in league with the French.

Edgeworth's thoughts again turned to politics because of his wish to promote popular education and because of his interest in the question of the proposed Union with Great Britain about which he was ambivalent. He became the member for St Johnstown in 1798 and took his seat in the last Irish Parliament.[30] He accepted intellectually a number of the arguments for the Union but voted against the proposal because he felt the majority in the country were against the proposed Union and because of the methods used by the Irish Administration.[31] He was active also on the question of education, was successful in having a select committee appointed and introduced an Education Bill that led to the Board of Education being established after the Union. On the demise of the Irish Parliament, he turned once again to his interests as a landlord, as an educator and as an inventor.

Experiments on Transport

Richard Lovell Edgeworth was interested throughout most of his life on road transport. Reference has already been made to the design of a phaeton through which he made the acquaintance of Erasmus Darwin. Edgeworth's other early experiments on transport involved a single-wheeled horse-drawn carriage, a velocipede and a carriage powered by a sail.

Following his final return to Ireland in 1782, Edgeworth continued to experiment on land transport and to publish the results of some of these experiments.[32-38] In 1788 he reported to the Royal Irish Academy on experiments on wheeled carriages; in 1801 a letter of his on a general system of Railroads appeared in *Nicholson's*

Journal; in 1806 another letter appeared in the same Journal on an odometer for a carriage; in 1813 he published an essay on Roads and Wheel Carriages; and during his final illness he published a further article on Wheel Carriages in *Nicholson's Journal* in 1816. The last of these publications was concerned with a series of experiments which he had planned many years before but only carried out for the Dublin Society in April 1815 and May 1816.

During a prolonged visit to Lyons in 1771, Edgeworth became seriously involved for the first and only time in water engineering.[39] He became concerned with the diversion of the Rhone to a new channel so as to facilitate an extension of the city of Lyons. Edgeworth devised a method of replacing the purely manual method of loading and unloading boats by a method involving a moving platform supported by piles. He records the damage done by a flood when the company took no action despite a ten-day warning. During the second winter, Edgeworth designed a number of flourmills which the company proposed to erect on some of the reclaimed land. Due to the death of his wife, he returned to England and did not remain to complete the project. His services were given gratuitously but the company involved conferred on him a plot of the reclaimed land which was later confiscated during the Revolution.

Experiments on Communication

Richard Lovell Edgeworth's interest in exploring telegraphic communication dated back to 1767 when he first met Sir Francis Delaval through a common interest in conjuring and through whom he was introduced to the world of fashionable London.[40] His interest in the topic resulted in a bet concerning the time required to convey the results of a race from Newmarket to London. Edgeworth proposed to use a telegraph system based on an idea of Robert Hooke but was honest enough to reveal this and the bet was called off. Edgeworth and Delaval carried out tests in London and on returning to Berkshire, Edgeworth continued to experiment on telegraphy over a distance of 16 miles, moving the setting of windmill arms in order to maintain secrecy.

After the rumours of a French invasion of Ireland in 1794, Edgeworth resumed experiments on telegraphy with a view to military application.[41] In August 1794, he constructed a few machines for conversation between Edgeworthstown and Packenham Hall, a distance of 12 miles and shortly afterwards exchanged messages between Ireland and Scotland. Between September and November, he carried out experiments and planned a vocabulary with the assistance of his schoolfellow and long-time friend John Foster, the

Speaker of the Irish House of Commons. A memorial was presented to Lord Camden the Lord Lieutenant early in 1795 but there was no response. The results of these experiments were reported in a memoir presented to the Royal Irish Academy.[42]

In September 1796, there were renewed rumours of a French invasion and Edgeworth wrote to Lord Carhampton the Commander in Chief and to Pelham, the Chief Secretary for Ireland and undertook in a one-page memorial to convey intelligence from Dublin to Cork and back to Dublin by means of a relay of 14 or 15 stations at an annual cost of £100 and at proportional cost for other distances. The initial response was favourable but a month later word was received from London that the matter would not be further pursued.

A few weeks later, the folly of this decision reached in London became apparent when a large French fleet arrived in Bantry Bay. Edgeworth sent an immediate message to the Chief Secretary offering to erect the telegraph machines which he had assembled on any line required in order to reconnoitre with portable telegraphs. No answer was received other than an acknowledgement. Despite his disappointment at these rebuffs, Edgeworth prepared a memorial to Lord Charlemont and published it as a pamphlet which is remarkably temperate and good humoured.[43] Again there was no official response and the telegraph was not available to the military or government during many widely separated disturbances of 1798.

Edgeworth returned to the development of a telegraph system in 1803 when requested to do so by the government. Copies of the letters between him and the government in Dublin Castle in 1803 and 1804 are preserved in twelve volumes in the manuscript division of the National Library.[44] Also in the National Library is a manuscript book of code for the telegraph containing a listing of words or phrases corresponding to 7,357 combinations of four figures indicated by the position of the points on four telegraph machines.[45] At the final test in 1804 in the presence of the Lord Lieutenant, messages and answers were transmitted between Dublin and Galway in 12 minutes.

CHARLES VALLANCEY, MILITARY ENGINEER

Outline of Career

Charles Vallancey is taken as an example of a military engineer in Ireland because more is known about him than about any of his colleagues. This is because of his leading role among Irish antiquarians in the second half of the 18th century.[46] Nevertheless, an examination of the primary sources reveals a good deal of information about his engineering activities. Charles Vallancey was born about 1725 apparently in Flanders of a family of French descent settled in England, brought back to England as a child and educated at Eton. Charles attended the Royal Military College and was commissioned as an Ensign in the Tenth Regiment of Foot in 1747, became a Colonel in 1782, a Major-General in 1793 and a Lt-General on the lst January 1798.

In 1750 Vallancey began to translate engineering works from the French and to undertake consulting work, apparently in order to earn money to support his family of ten children. His first translations were of French publications on military engineering. These were followed by works on civil engineering including the substantial Treatise on Inland Navigation based on the translation and reorganisation of the works of the Italian experts.[47] This may have led him to study further the questions of river and canal engineering since a few years later we find him acting in a consulting capacity in relation to the Boyne and the Grand Canal.[48] Also of note is a translation from the French of de la Rue's treatise on stone cutting apparently written in connection with his design and supervision of the new Queen's Bridge at Arran Quay.[49] In the 1780s, Vallancey was largely engaged in work on fortifications in Dublin, Cork and Carrickfergus. His work on the fortification of Cork Harbour in the 1790s may have influenced the French fleet to rendezvous at Bantry Bay in 1796 rather than attacking Cork directly.

In the 1760s, if not earlier, Vallancey began to take an interest in Irish antiquities and in the Irish language. He was elected a member of the Dublin Society in 1763 and was responsible for the formation of a Committee of Antiquarians in 1772. He was a founder member of the Royal Irish Academy in 1785 and one of the seven original members of the Antiquities Committee of its Council.[50] His views on the origins of the Irish, which he advocated with great vigour, were typical of his period but quite erroneous. However, his contribution to the promotion of archaeological studies and the conservation of Irish antiquities should not be underestimated.

We can gain an insight into the wide interests of Vallancey because a copy of the auctioneers list of 1417 items from his library is among the pamphlet collection of the National Library.[51] These are grouped into manuscripts and notebooks (183) Natural History (48), Mathematics (98), English History (112), Poetry and Plays (95), Dictionaries and Grammar (126), Voyages and Travel (62), Latin miscellaneous (40), General History (83), Classics (130), Irish History (194), Manuscripts (26), Northern History and Antiquities (83), and books of Prints etc. (42). Some of the books relating to military engineering and civil engineering will be referred to later.

Work in Military Engineering

The history of military engineering in Ireland has not been studied in any detail. As might be expected, the majority of those listed in the *Biographical Dictionary of Engineers in Ireland 1600–1730* are military engineers.[52] Some information is available in general histories of the Royal engineers and in short contributions to the *Royal Engineers Journal* but no clear picture emerges.[53-54] In 1798 itself the engineer officers listed for Ireland were Charles Vallancey, Director and Major General together with Colonel Charles Tarrant, Lt-Colonel James Farrier, Major John Brown, Captain Alexander Taylor and Captain Henry Eustace, First Lieutenant George Armitage and Ensign Danielle Corneille.[55] Vallancey's main activities as a military engineer seem to have been in mapping and in defence works.[56-62] A map of Gibraltar in the British Museum was executed and signed by him and has been described as workman-like.[56-57] He was posted to Cork where he spent most of the 1750s as an extra engineer. In 1762 he was promoted to Engineer in Ordinary and resided for most of the next two decades in Dublin.

New opportunities for professional work came to Vallancey with the appointment of his Eton contemporary Lord Townshend as Lord Lieutenant of Ireland in 1767. Townshend composed an elaborate despatch on the Defence of Ireland for which Vallancey prepared the illustrative maps. In delivering the Townshend dispatch to the authorities in London, Vallancey, besides complaining that he was 'starving with a large family', proposed a plan for a one-inch map of Ireland to be carried out by the Corps of Engineers in Ireland at double pay. Such a project was commenced under Vallancey's direction in 1776 and copies of the maps are available in a number of manuscript collections.[57-59]

Between 1790 and 1796 Vallancey was in command of the forts in Cork Harbour which was a responsible post since the French were expected to invade and to land at one of the southern ports. Vallancey was acquainted with the classic works of the period on fortification as evident from the contents of his library which included a 1762 edition of Vauban's *New Method of Fortification*, Carmantaigne's *Architecture Militaíre* of 1741, and Müller's *Practical Fortification* of 1774. He made many improvements to the defences of Cork Harbour. Two manuscript reports have survived, one on the Defensive State of Cork Harbour and the other on the defence of the Southern Coast.[60-62] The latter report concentrates on the defence of Waterford, Cork and Limerick and stresses the need for providing depots for transport and arrangements for the victualling of towns as well as the improvement and maintenance of fortification.

Work in Civil Engineering

Vallancey was no mere translator of works on civil engineering. His *Treatise of Inland Navigation* of 1763 is a well-organised treatment of the subject. There is evidence that Vallancey possessed engineering judgment as well as paper knowledge. In 1771 he reported on the alternative Grand Canal proposals and recommended the adoption of the Southern line in opposition to the views of John Traill who was the engineer to the Grand Canal Company.[63] John Smeaton the leading civil engineering consultant of the day was called upon to adjudicate on the question and found in favour of Vallancey.[64]

As in the case of military engineering, the listing of Vallancey's library reveals the sound basis of his knowledge of hydraulic engineering. Among his books were an Italian translation of Vitruvius dated 1515 as well as an English translation published in 1791, *Traite de Moyens de Rendre les Rivières Navigable* published in Amsterdam in 1696, Switzer's *System of Water and Water Works* published in 1734 and the 1737 edition of Belidor's masterpiece on *Architecture Hydraulique.*

Vallancey also made contributions in structural engineering. He was thanked by the Mayor, Sheriffs and Grand Jury of Kilkenny City for his great skill, expedition and readiness in erecting temporary bridges over the Nore in 1763 when two permanent bridges were carried away by floodwater.[65] He designed the Queen's Bridge near Arran Quay to replace Arran Bridge also destroyed by floods in 1763. The new bridge to his design was built between 1764 and 1768 and has received much favourable comment in the following two centuries.[66] He was responsible for the completion of the construction, but not the design, of the old pier at Dun Laoghaire built in the early 1760s.

Finally in the last few years of his life, he acted as Chairman of the Bogs Commission of 1807-1812 whose pioneering reports provide a wealth of information which proved of great value in later years. One of the nine engineers engaged in this task was Richard Lovell Edgeworth who was assisted by his son William then embarking on a career as a civil engineer.[67-68] Richard Lovell Edgeworth was at that time 65 years of age and in poor health.[69] Charles Vallancey was over 80.

WILLIAM CHAPMAN, CIVIL ENGINEER

Family and early life

William Chapman (1749–1832) was first of the new breed of professional civil engineers to work in Ireland on a full time basis.[70] Though his period of full time work in Ireland on canal engineering occupied only eleven years (1783–1794), he was associated with the Grand Canal Company as a consultant for a further five years and supervised the building of the Ringsend Docks. He came from a well-to-do Quaker family who were freemen of Newcastle-upon-Tyne. His father, Captain William Chapman, was a shipowner with an interest in engineering and in science who published two papers in the *Philosophical Transactions of the Royal Society*, one on the desalination of sea-water and the other on the discovery of a Jurassic fossil of an unknown type of crocodile which was named after him, *Teleosaurus chapmanni*. The younger William Chapman was born in Whitby in 1749 and received a liberal education at different public schools. As a young man he was given command of a merchant vessel trading with western and northern Europe for a period of three years. In the three years after this he studied numerous harbours in Britain and on the continent and developed an amateur interest in harbour engineering. In 1778, he and his brother John took a lease of Byker colliery and Wallsend colliery in the Newcastle district. In this connection he became acquainted with James Watt whose firm of Boulton and Watt provided a steam engine for pumping at the Byker colliery. He corresponded with Watt on matters relating to steam engines other than those relating to the Byker pumping engine and also developed an improved method of coal-winding. In 1779 he won second prize in an open competition for designs to improve the pumping engine at Rotterdam. Though all seemed to go well at Byker, the sinking of the first shaft at Wallsend led to financial difficulties and the mortgage on the venture was foreclosed. William and John Chapman were declared bankrupt in 1782.

During this period of 1770s William Chapman sought to satisfy his interest in engineering by becoming acquainted with persons of an older generation such as his father's friend John Wooler who was harbour master of Whitby, James Watt and his partner Matthew Boulton. These persons became his friends and all of them urged him to convert his amateur interest into a professional career. Following the collapse of the colliery venture, William Chapman decided to follow this advice and considered trying his fortune in Ireland. Late in 1783, William Chapman , then aged 34, accompanied

Matthew Boulton to Ireland and, apart from short trips home, spent the next eleven years in Ireland.

Canal Construction in Ireland

At first, Chapman was employed as an agent for Boulton and Watt supervising the erection of an engine for a colliery in County Tyrone and other minor activities. Meanwhile he was seeking employment of more consequence, preferably of a civil engineering nature. He had no real success until he wrote a prize essay on the effects of the River Dodder on Dublin Bay. A manuscript map which was part of the essay survives in the manuscript division of the National Library.[71] Shortly afterwards he published a report advocating that on environmental considerations the proposal to link the terminus of the Grand Canal at James Street Basin by a short cut to the River Liffey should be replaced by a circular line with eight locks over a four mile length to join the Liffey in a dock opposite the new Custom House.[72] His suggested plan was carried out in 1790–1792.

William Chapman's first substantial employment in Ireland was as engineer with a salary of £200 per annum on the Kildare Canal (1787–1789). This was a 2¹/₂ mile cut from the Grand Canal near Sallins to Naas with a rise of 44 feet in 5 locks and involving bridges at Osbertstown, Oldtown and Naas.[73] The work on this canal was notable for the building of the first modern skew bridge in these islands. Prior to this, bridges carrying roads at an angle to a river or canal were constructed in such a way as to be either dangerous in design or wasteful of material. The concept and means of setting out based on a cylindrical helix to obviate these defects are described by Chapman himself in a four page article in the 1813 edition of *Rees' Cyclopaedia*.[74]

From this time onwards, William Chapman's professional engagements increased. A number of these were connected with the Grand Canal for which William Jessop was the consulting engineer. These included (a) the survey between 1789 and 1792 of the entire length of 53 miles westward from the summit level near Robertstown to Shannon Harbour; (b) supervising and reporting in 1791 on the progress of works crossing the Bog of Allen; (c) in 1793 and 1794 rebuilding sixteen of the original eighteen locks on the eastern section between Dublin and the summit level near Robertstown; and (d) the design of several new bridges. In the light of difficulties encountered in boggy ground near Robertstown, Chapman took special precautions in surveying the bogs near Tullamore.[75] He cut out from different strata of the exposed bog cylindrical specimens of peat and left them on a marble mantelpiece in winter for three or four months. On the basis of the loss of weight, the total subsidence was estimated and line was chosen. Chapman also assisted William Jessop in the early stages of the design and construction of Ringsend Docks (1792–1796) consisting of two linked wet docks and three wet docks. His brother Edward Walton Chapman acted as resident engineer on this project.

William Chapman also prepared a number of reports on other schemes of inland navigation. These included (a) proposals for improvement of the Barrow Navigation between St Mullins and Athy by replacing four of the existing locks and building ten new locks to accommodate eighty ton boats; an inland cut from Burns Island to Leighlin in Carlow; the navigation of the River Shannon from Lough Allen to Killaloe; a proposed canal from Thomastown to Kilkenny; a proposal to improve Arklow Harbour and to make the Avoca navigable; improvement of navigation on the Erne from Woodford Lough to Ballyshannon.[76-81] Works were only commenced on three of these seven proposals namely the Lower Barrow, the Shannon from Lough Derg to Limerick and Lough Erne to Ballyshannon.

Later career as a consultant

By 1794, William Chapman was well established as a civil engineer in Ireland. This was recognised by his election as a member of the Royal Irish Academy in that year. However, despite his future career in Ireland being assured, he chose to return to his own country and his own family. The first work he was engaged on after his return was a series of reports on the proposed canal between the Tyne and Solway Firth which became the subject of a violent parliamentary controversy between conflicting interests.

In the following years (1796–1803), William Chapman was concerned with large scale drainage and flood protection schemes in east Yorkshire and with harbour works. At this time he published his *Treatise on the Various Systems of Inland Navigation* which deals largely with the comparison of navigation locks and of inclined planes as used by Ducart and advocated by Fulton and others.[82] In 1806 Chapman's career underwent another change at a time when canal construction was coming to an end and railway construction had not begun. Under these circumstances, Chapman's interest turned again to coal mining and he developed a number of improved services for which he received patents. After 1816 his civil engineering practice again expanded.

In 1796, William Chapman became a number of the Smeatonian Society of Civil Engineers on the proposal of William Jessop, seconded by John Rennie and Robert Mylne. He never became a member of the Institution of Civil Engineers but a few years after his death his wife donated an extensive set of his published reports to the Institution. These are still available in the Institution Archives.

BRITISH CONSULTANTS IN IRELAND

John Smeaton (1773–1775)

Though John Smeaton's contact with Ireland between 1770 and 1775 is a tiny part of a long and varied career, it is of interest because of his position as the first professional civil engineer and of his influence on engineering practice.[83] John Smeaton (1724–1792) was born at Austhorpe near Leeds, the son of an attorney, and as a young boy was interested in mechanical pursuits. He was educated at Leeds Grammar School and entered his father's office in his sixteenth year. He went to London in 1742 but soon abandoned his legal studies to enter the employment of an instrument maker. He opened a shop of his own in 1750.

At this time, he regularly attended the meetings of the Royal Society and was elected a fellow in 1753. One of the eighteen papers which he presented to the Royal Society was the intriguing account of *An Experimental Enquiry concerning the Natural Powers of Wind and Water to Turn Mills*[84] which won him the prestigious Copley Medal of the Society in 1759. In 1755 he was recommended by the President of the Royal Society as the most suitable person to design the replacement for the Eddystone Lighthouse. The outstanding success of this project which was completed in 1759 launched him on a career that made him the most eminent engineer of the 18th century in Britain.[85]

Smeaton established a scheme of organisation that distinguished between the functions of the consulting engineer and the resident engineer. He built a four-storey tower in the grounds of his house at Austhorpe near Leeds and used it as an office, workshop and study. His relaxation was based on astronomical and antiquarian pursuits. He contributed several papers on astronomical topics to the Royal Society between 1768 and 1788. After his death, Sir Joseph Banks, President of the Royal Society and a member of the Society of Civil Engineers, purchased all of his manuscripts, reports and drawings. The entire set of reports were published by the Smeatonian Society of Civil Engineers in three large volumes.[86] Shortly afterwards his eighteen papers in the *Philosophical Transactions of the Royal Society* were also reprinted.[87]

Reference has already been made to the differences of opinion between Charles Vallancey and John Traill concerning the line of the Grand Canal and the junctions with the Barrow, Boyne and Shannon.[88] In June 1771, one of the subscribers to the proposed company Redmond Morres wrote to John Smeaton asking him to come to Ireland for a month or six weeks to examine the ground and advise on the matter. Smeaton, who was then engaged on the Forth and Clyde Canal, replied at the beginning of August that it would be 'absolutely impossible to take the time necessary to examine so extensive an affair'. There followed a protracted correspondence which was printed as a 59 page volume.[89] Smeaton, accompanied by Jessop, finally came to Ireland in June 1773 and spent almost two months examining the line with John Traill. Later in the year, Smeaton produced a report with specific recommendations on the line of the canal, the economic disadvantages in combining canal construction and bog drainage, the size and type of boats, and the design of locks.[90]

Smeaton had also been asked to advise on the Tyrone Navigation and on the design of the inclined planes on Ducart's Canal. He did not travel to the area but sent William Jessop to report. After receipt of Jessop's findings, John Smeaton wrote a report on the matter in April 1774 in which he criticised the design and the works as carried out but recommended that in view of the amount already spent on the project the works be completed by constructing the inclined planes.[91] This visit of Jessop to Ireland was the first of many as he returned later to advise the Grand Canal Company over a period of two decades.

William Jessop (1773-1806)

William Jessop was born in Plymouth in 1745, his father Josias being a foreman shipwright in Plymouth Dockyard. Josias later acted as a superintendent of the construction of the Eddystone Lighthouse under John Smeaton from 1756-1759. William became a pupil of Smeaton's in 1759, and Smeaton became his guardian after his father's death in 1761. William Jessop after serving his time, remained on as an assistant to Smeaton from 1768 to 1774. In the latter capacity he accompanied Smeaton to Ireland in 1773. After starting to practise on his own account, he was concerned with a wide variety of projects mostly on inland navigation, harbours and docks.[92-93]

The 1773 visit of Jessop to Ireland as assistant to Smeaton has already been noted. The next record of visits to Ireland is for 1785 and 1789 with possible visits in the intervening period for which the Grand Canal records are missing.[94] From 1789 until 1802, he was clearly the main consultant to the Grand Canal Company of Ireland and his reports were entered in full in the Minutes of the Company.[95]

Between 1789 and 1800, William Jessop acted as consulting engineer on the extension of the Grand Canal from the summit at Robertstown to Tullamore with James Oates and John Killaly as resident engineers. In a report of July 1790 he described the results of his survey of this stretch where the key problem was the building of the canal in deep bog. It would appear that in early years insufficient time was allowed for subsidence but by 1794 Jessop seems to be in full agreement with Chapman on the need for continual deepening of the drains and for allowance between drainings of adequate time for settlement. The final form of technique is described in the description of the work on the Ballinasloe Canal by Bernard Mullins and his son M.B. Mullins.[96]

In a later report of September 1796, Jessop deals at length with the problems of the Robertstown-Tullamore line, particularly that of the Edenderry embankment. In the following year he travelled to Ireland in December and reported on the Edenderry breach that had occurred in the previous August. In 1796 and 1797 he also reported his views on the appropriate line from Tullamore to the Shannon based on his own survey and the previous recommendation by Chapman which was based on a survey carried out with the assistance of Killaly and Oates.

In 1790, the Directors of the Grand Canal Company asked Jessop to report on their difficulties in relation to the proposed Royal Canal and on the Circular Line at Dublin as well as on the Robertstown to Tullamore stretch. In 1793, Jessop was consulted by the Royal Canal Company on four points and reported on three of them.[97]

William Chapman had in 1785 proposed and surveyed a circular line from James Street Harbour to Ringsend. In response to the Director's request in 1790, Jessop supported the proposal and suggested some minor modifications. The work on the circular link was carried out between 1790 and 1793 with Jessop as consulting engineer, assisted by Chapman, and with Archibald Miller as resident. The works on the Ringsend Docks were carried out between 1792 and 1796 with Edward Chapman, brother of William, as resident engineer. The opening ceremony for the new docks was performed with great pomp and entertainment by Lord Camden, the Lord Lieutenant and reported in detail by the Dublin papers.

In April 1800, Jessop visited Ireland at the urgent request of Sir John Macartney, Chairman of the Grand Canal Company and Sir Richard Griffith another shareholder who between them held a controlling interest. Their request was to survey the line of a proposed ship canal from Ringsend Docks to Dun Laoghaire and to report on the practicality of making a harbour for large vessels at Dun Laoghaire.[98]

In 1794, Jessop reported for the Limerick Navigation Company on the state of the middle Shannon which had been surveyed by Chapman in 1791 and reported on by William Chapman and John Brownrigg for the Grand Canal Company. Jessop's 1794 report proposed a comprehensive development with a navigation trackway to facilitate the conveyance of coal from Lough Allen and of agricultural produce from the counties bordering the river to provide trade for the Grand Canal when completed. In November and December 1801, Jessop spent 29 days in Ireland in order to update his 1794 Report.

Jessop's last visit to Ireland was in 1806 on behalf of the Barrow Navigation Company in relation to the improvement in navigation between the branch of the Grand Canal at Athy southward to Carlow and to the tidal estuary at St Mullins. This work was never carried out.

John Rennie (1790-1821)

John Rennie was born near Edinburgh in June 1761, the son of a farmer.[99] From the age of five or six, John Rennie took an interest in mechanical things and later spent much time with the local millwright Andrew Meikle who is remembered for having patented the spring sail for windmills in 1772 and the first threshing machine in 1788. When Rennie completed his studies at the village school at the age of twelve, he spent two years training with Meikle in the crafts of the carpenter, the blacksmith and the stone mason. At his own request he was sent at the age of fourteen to Dunbar High School where he became head boy and excelled in all subjects, showing particular brilliance in mathematics as reported by a visitor to the school.[100] After two years, he returned home and combined practical work under Andrew Meikle with private studies. In 1779 he set up on his own at the age of eighteen and progressed from dealing with the overflow from Meikle's business to projects under his own responsibility.

In 1780 he went up to Edinburgh University where he spent three years studying chemistry under Professor Joseph Black (1728-1799) and natural philosophy under Professor John Robison (1739-1809), both outstanding scientists of their generation. After completing his studies in 1783, Rennie made his own Grand Tour of the leading engineering works then in progress in England.

In 1784, John Rennie received an offer from James Watt to enter the employment of Boulton and Watt to work on the large-scale project at Albion Mills near Blackfriars Bridge in London. He abandoned his own small but flourishing business and accepted the offer. While supervising the work at Albion Mills between 1784 and 1788, he worked on designs of machinery for a number of large-scale mills in Britain and on the Continent. About 1790, Rennie became more active in hydraulic engineering than mechanical engineering. He was consulted about a proposed canal from Bishop's Stopforth to King's Lynn and about the drainage of 50,000 acres at the Bottesham and Soham Fen.

John Rennie's work in Ireland at this time was largely in connection with the first Custom House Dock in Dublin.[101-102] The plans for the new Custom House included the concept of associated docks surrounded by bonded warehouses and connected to the River Liffey by locks. John Rennie was appointed consulting engineer for this part of the project. The first of these docks was opened in 1796 but without any of the ceremony and fanfare that accompanied the opening in the same year of Ringsend Dock on the opposite side of the Liffey. In the last decade of the 18th century, John Rennie also acted as a consultant to the Royal Canal Company. In a report dated 1802, he proposed alterations in the proposed line of the Royal Canal from Dublin to the Shannon near Longford. It is recorded (Boucher 1967, p.74) that one of his recommendations to the Company was to get the resident engineer Townsend a good comfortable house since he would be much more efficient if he were made comfortable.[103]

NOTES

1. Charles Joseph Singer, Eric John Holmyard, Alfred Rupert Hall and Trevor I. Williams, *A History of Technology. Volume 3: From the Renaissance to the Industrial Revolution 1500–1750* (8 volumes, Oxford: University Press, 1957); Chapter 14 (pp. 347–76) 'Military technology' by A.R. Hall, Chapter 17 (pp. 438–70) 'Canals and River Navigation before 1750' by A.W. Skempton.
2. Frederick Binkerd Artz, *The Development of Technical Education in France 1500-1850* (Cambridge: M.I.T. Press, 1966).
3. Charles Stewart Gillmor, *Coulomb and the Evolution of Physics and Engineering* (Princeton University Press, 1971).
4. Nancy J. Curtin, *The United Irishmen: Popular Politics in Ulster and Dublin 1791–1798* (Oxford: Clarendon Press, 1994), pp. 176–81.
5. John Keane, *Tom Paine: a Political Life* (London: Bloomsbury Publishing, 1995). See also D.H. Hooke, *Paine* (New York: Norton, 1974), pp. 160–87, 347–8 and 368–9.
6. John Smeaton, *Journal of a Five Weeks Tour in Flanders and Holland in the Year 1755 Trinity House, London. MS 114* edited by A. Titley and published as Newcomen Society Extra Publication No 4 (London, 1938).

7. Anon., Introduction to Volume 1 of *Transactions of the Institution of Civil Engineers.* (London: J. Weale, First Edition, 1836, Second Edition 1842), pp. iii–xxvii.
8. A.W. Skempton, 'Engineers of the English river navigations 1620-1760', *Transactions of the Newcomen Society,* 29 (1955): pp. 25-54.
9. Richard Castle, *Essay on Inland Navigation.* National Library, Ms. 2737. Photocopy in Irish Architectural Archive RP.B. 17.9, (1730). Reproduced and edited by J.H. Farrington. *Transport History,* 5 (1972): pp. 67-89, 155-67.
10. *Irish Commons Journal,* Reference to Newry Navigation, volume 5, appendix (1750) pp. 26-8; p. 50.
11. Richard Castle, *An Essay towards supplying the city of Dublin with Water.* Printed by S. Pepy (1735), 32 pp. Royal Irish Academy Haliday Pamphlets 121 no. 8, Irish Architectural Archive Pamphlets CS. ICS. 1 (7). See also G. Stokes, *Observations on a late Essay by Mr. Richard Castle* (1735). Royal Irish Academy Haliday Pamplets 121 no. 7, 6 pp. Copy in Irish Architectural Archives 91/83.
12. William Alan McCutcheon, *The Canals of the North of Ireland* (Dawlish: David and Charles, 1965).
13. Anon., Richard Castle *Anthologia Hibernia* (October 1793), pp. 242-3.
14. Knight of Glin, 'Richard Castle, architect. His biography and works. A synopsis', *Irish Georgian Society,* 7(1): pp. 31-8.
15. M.B. Mullins, Presidential Address to the Institution of Civil Engineers of Ireland. *Transactions of the Institute of Civil Engineers of Ireland,* 6 (1859): pp. 1-186.
16. Knight of Glin, 'David Duckart. Last of the Palladians', *Country Life,* CXLII (Sept. 28 and Oct. 5 1967): pp. 735-9 and 798-801.
17. William Alan McCutcheon, *Industrial Archaeology of Northern Ireland* (Department of the Environment: HMSO, Belfast).
18. *Irish Commons Journal*: References to Tyrone Navigation (1750-1800) 9 (1753): p. 94; 11 (1759): pp. 932-7; 12 (1761): pp. 482-4; 8(1767): pp. 179-80; 7(2) (1768): pp. 249-50; 23 (1785): p. 546; 24 (1787): p. 173; 25 (1788): p. 182; 19(2) (1800): p. 1708.
19. D.H. Tew, 'Canal lifts and inclines with particular reference to those in the British Isles', *Transactions of the Newcomen Society,* 28 (1951): pp. 35-58.
20. *Irish Commons Journal,* 7 (1767): pp. 179-80. 17 November 1767. Proposals by Omer and Ducart for Tyrone Navigation.
21. F. Sloane, Map of Ducart's Canal. Royal Irish Academy. Map Recess. Case 1, Shelf A.
22. John Smeaton, *Reports* Society of Engineers, volume 2, pp. 247-77. *Reports on Grand Canal* (1773, 1775); volume 2, pp. 278-9. *Report on Tyrone Navigation* (1774), (London, Longman, Hurst, Rees, Orme and Brown, 1812). Volume 4 in the Library of Institution of Engineers of Ireland.
23. Ronald C. Cox and Michael H. Gould, *Civil Engineering Heritage: Ireland* (London: Telford Publications, 1998).
24. Richard Lovell Edgeworth, *Manuscript of draft memoirs (1808-1809).* Basis of M. Edgeworth, *Memoirs of Richard Lovell Edgeworth* (2 volumes, London: Hunter and Craddock, 1820), volume 1. National Library, Ms.

7362.
25. M. Edgeworth, *Memoirs of Richard Lovell Edgeworth* (2 volumes, London: Hunter and Craddock, 1820). Reprinted in facsimile (2 volumes, Shannon: Irish University Press, 1969), volume 1: p. 25.
26. Desmond John Clark, *The Ingenious Mr. Edgeworth* (London: Oldbourne, 1965).
27. Elizabeth Kowaleski-Wallace, *Their Father's Daughters: Hannah More, Maria Edgeworth, and patriarchal complicity* (Oxford University Press, 1991).
28. Maria Edgeworth, *Memoirs of Richard Lovell Edgeworth* (2 volumes, London: Hunter and Craddock, 1820). Reprinted in facsimile (Shannon: Irish University Press, 1969), volume 2, pp. 205–13.
29. Note 28, pp. 214–35.
30. Note 28, pp. 242–9.
31. Richard Lovell Edgeworth, *The Substance of Three Speeches delivered in the House of Commons of Ireland, February 6, March 4 and March 21 1800 upon the Subject of an Union with Great Britain* (Printed by J. Johnson, 1800).
32. R.L. Edgeworth, 'Experiments on the resistance of air,' *Philosophical Transactions of the Royal Society*, 73 (1783): pp. 136–43.
33. R.L. Edgeworth, 'An account of some experiments on wheel carriages', *Transactions of the Royal Irish Academy*, 2(1) (1788): pp. 73–80.
34. R.L. Edgeworth, 'Letter on the practicability and advantages of a general system of rail roads', *Nicholson's Journal*, 1 (1801): pp. 221–23.
35. R.L. Edgeworth, 'Letter containing a description of an Odometer for a Carriage and some remarks on the Patent Boring Machine of Mr. Ryan', *Nicholson's Journal*, 15 (1806): pp. 80–3.
36. Richard Lovell Edgeworth, *An Essay on the Construction of Roads* (Printed by J. Johnson 1813).
37. R.L. Edgeworth, 'On aerial navigation', *Nicholson's Journal*, 47 (1816).
38. R.L. Edgeworth, 'On wheel-carriages', *Nicholson's Journal*, 48 (1816).
39. See note 25, pp. 257–72 and 315–22.
40. See note 25, pp. 140–9.
41. See note 28, pp. 154–72.
42. R.L. Edgeworth, 'An essay on the art of conveying secret and swift intelligence', *Transaction of the Royal Irish Academy*, 6(1) (1795): pp. 95–139.
43. Richard Lovell Edgeworth, *A Letter to the Rt. Hon. the Earl of Charlemont on the Tellograph and on the Defense of Ireland.* (Dublin: Printed by P. Byrne, 1797).
44. Richard Lovell Edgeworth, *Manuscript correspondence with Dublin Castle regarding the Telegraph (1801–1811).* National Library, Ms. 8182: 1–12.
45. Richard Lovell Edgeworth, *Vocabulary Complete Manuscript book of Code for the Telegraph*, (1803). National Library, Ms. 7393.
46. M. Nevin, 'General Charles Vallancey 1725–1812', *Journal of the Royal Society of Antiquaries of Ireland*, 123 (1993): pp. 19–58.
47. Charles Vallancey, *A Treatise on Inland Navigation* (Dublin: G. & A. Ewing, 1763).
48. Charles Vallancey, *A Report on the Grand Canal, or the Southern Line*

(Dublin: Printed by T. Dyton, 1771), 66 pages. National Library. See also *Extract from a Report on the state of the Boyne Navigation*, National Library, Ms. 7352.

49. Charles Vallancey, *Stone cutting; Translation from the French of de la Rue* (Dublin: Thomas Ewing, 1776).
50. Tarlach O Raifeartaigh, (editor) *Royal Irish Academy: a bicentennial history* (Dublin: Royal Irish Academy, 1985).
51. Thomas Jones, *Catalogue of a Valuable collection of Books, Manuscripts and Irish History, the Library of the late celebrated Irish Historian General Charles Vallancey* (Dublin: Thomas Jones, Auctioneer, 6 Eustace Street, 1813).
52. R. Loeber, 'Biographical Dictionary of Engineers in Ireland 1600-1730', *Irish Sword*, 13 (1977, 1978, 1979): pp. 30-44, pp. 106-22, pp. 230-55, pp. 283-314.
53. Whitworth Porter, *History of the Corps of Royal Engineers* (London: 1889).
54. J.J. Crooks, The Corps of Engineers, Ireland, *Royal Engineers Journal*, 23 (1916): pp. 12-5.
55. *A List of the Officers of the Army and Marines* (London: War Office, 1798), p. 372 of Corp of Royal Engineers in Ireland.
56. Charles Vallancey, *Plan of Gibraltar*. British Library, Ms. Add. 21576. 3B F3h (1750)
57. J.H. Andrews, 'Charles Vallancey and the Map of Ireland', *Geographical Journal* 132(1) (1966): pp. 48-61, and J.H. Andrews, 'Charles Vallancey's Maps of Ireland: a new reference', *Geographical Journal*, 146 (1980): p. 150.
58. Charles Vallancey, *Survey of 1776*. Royal Irish Academy, Ms. 2737.
59. Charles Vallancey, *An Essay on Military Surveys*. British Museum, Map Room K. Top LI 31-2. 6 Tab 64. (1779).
60. Charles Vallancey, *Defensive State of Cork Harbour*. British Library, Ms. Add. 33.119; Add. 33.105. (1797).
61. Charles Vallancey, *The Defence of the Southern Part of Ireland*. Copy in Vallancey's own hand, (1797), Archives of the Catholic Diocese of Ferns, Wexford.
62. M. Nevin, 'The defence of the southern part of Ireland by General Vallancey, Chief Engineer', *Journal of Royal Society of Antiquaries of Ireland*, 125 (1995): pp. 5-9.
63. R. Delany, 'John Trail, Grand Canal Engineer', *Journal of the County Kildare Archaeological Society*, 14(5) (1970).
64. Ruth Delany, *The Grand Canal of Ireland* (Newton Abbott: David and Charles, 1973), pp. 18-9.
65. Note 46, p. 25.
66. Maurice Craig, *Dublin 1660-1860*. (1952). Reference on p. 288 to Queen's Bridge designed by C. Vallancey.
67. Richard Lovell Edgeworth, Report on District No. 7. *Second Report of the Commissioners appointed to inquire into the Nature and Extent of the Bogs in Ireland* (1810), Appendix No. 8, pp. 179-204.
68. W. Edgeworth, 'On the triangulation of Roscommon', *Transactions of the Royal Irish Academy*, 14(1) (1823): pp. 63-7.
69. Note 28, pp. 313-20.

70. Anon., 'Memoir of William Chapman', *Weales Quarterly Papers,* 1(1) (1843): pp. 1–28.
71. William Chapman, Manuscript Map: *Plan of the River Dodder and of the proposed Channels to discharge it on the South Strand* Royal Irish Academy 16 G 14(22) April 1785.
72. William Chapman, *Observations on the advantages of bringing the Grand Canal round by the Circular Road into the River Liffey.* Report to the Wide Street Commissioners. (Dublin: Printed by P. Byrne, 1785).
73. R. Delany 'The County of Kildare Canal', *Journal of County Kildare Archaeological Society,* 15(1973): pp. 122–135.
74. W. Chapman, 'Oblique bridges', *Rees' Cyclopedia,* (1813), 4 pp.
75. See note 70, pp. 5–6.
76. William Chapman, *Report on Means of Perfecting the Navigation of the River Barrow from St. Mullins to Athy* (Dublin: Printed by W. Sleater, 1789).
77. William Chapman, *Remarks on the Proposed Plan of an Inland Cut from Burn's Island to Leighlin* (Carlow, 17th May 1791).
78. William Chapman, *Report on the Navigation of the River Shannon from Lough Allen to Killalow* (Limerick: Printed by A. Watson, for Shannon Navigation Company, 1791).
79. William Chapman, *Report on the Practicability and Expense of making a Navigable Canal from Thomastown to Kilkenny* (Kilkenny: Catherine Finn, 1792).
80. William Chapman, *Report on the Improvement of the Harbour of Arklow and the Practicability of a Navigation thence ... of the Avoca* (Dublin: Printed by J. Chambers, 1792).
81. William Chapman, *Report on the Means of making the Woodford River Navigable from Lough Erne to Woodford Lough and an Off-branch from the Lough Erne and Ballyshannon Navigation* (Limerick: Printed by A. Watson, 1793).
82. William Chapman, *Treatise on Various Systems of Canal Navigation* (London: I. & J. Taylor, 1797).
83. Alec Westley Skempton (ed.), *John Smeaton FRS* (London: Thomas Telford, 1981).
84. J. Smeaton, 'An experimental enquiry concerning the natural paower of water and wind to turn mills and other machines depending on a circular motion', *Philosophical Transactions,* 51 (1759): pp. 100–74.
85. John Smeaton, *A Narrative of the Building and a Description of the Construction of the Edystone Lighthouse* (London: Printed for the author, 1791).
86. John Smeaton, *Reports of the late John Smeaton F.R.S.* (3 volumes, London: Hurst, Rees, Orme, and Brown, 1812, Second Edition 1837).
87. J. Smeaton, *Miscellaneous Papers of John Smeaton, Civil Engineer, FRS comprising his communications to the Royal Society* (London: Longman, Hurst, Rees, Orme and Brown, 1814).
88. Note 64, pp. 18–9.
89. R. Morres and John Smeaton, *The Queries proposed by the Company of Undertakers of the Grand Canal, to John Smeaton, Esq. Engineer and F.R.S. Together with his General Report thereon* (Dublin: O. Nelson, 1773).

90. John Smeaton, *Reports 1773 and 1775. Reports on the Grand Canal.* (London: Longman, Hurst, Rees, Orme and Brown, 1812), volume 2, pp. 247–77.
91. John Smeaton, *Reports 1774. Report on the Tyrone Navigation.* (1812), Volume 2, pp. 278–79.
92. S. Hughes, 'Memoir of William Jessop', *Weales Quarterly Papers,* 1(2) (1844).
93. Charles Hadfield and Alec Westley Skempton, *William Jessop, Engineer* (Newton Abbot: David and Charles, 1979).
94. Note 93, p. 86.
95. Note 64, pp. 36–7
96. B. Mullins and M.B. Mullins, 'On the origin and reclamation of peat bog', *Transactions of the Institute of Civil Engineers of Ireland,* 2 (1846): pp. 1–48.
97. *Irish Commons Journal* (1800) 19(2) Appendix. Reports from Committee on Inland Navigation. First Report (23 June): pp. 1004–42.
98. William Jessop, *Report on Dunleary Harbour.* National Library P 131 JP 854 (1800).
99. Cyril Thomas Goodman Boucher, *John Rennie 1761–1821* (University of Manchester Press, 1963).
100. David Loch, *A tour through most of the trading towns and villages of Scotland: containing notes and observations concerning the trade, manufactures, improvements, &c. of these towns and villages* (Edinburgh, 1778). Quoted in Boucher (note 99), pp. 4–5.
101. Joseph Robins, *Custom House People* (Dublin: Institute of Public Administration, 1993), pp. 20–2.
102. John Rennie, Manuscript, letters and reports on Royal Canal. Institution of Civil Engineers Archives, London.
103. See note 99.

Index of people, places, institutions and societies